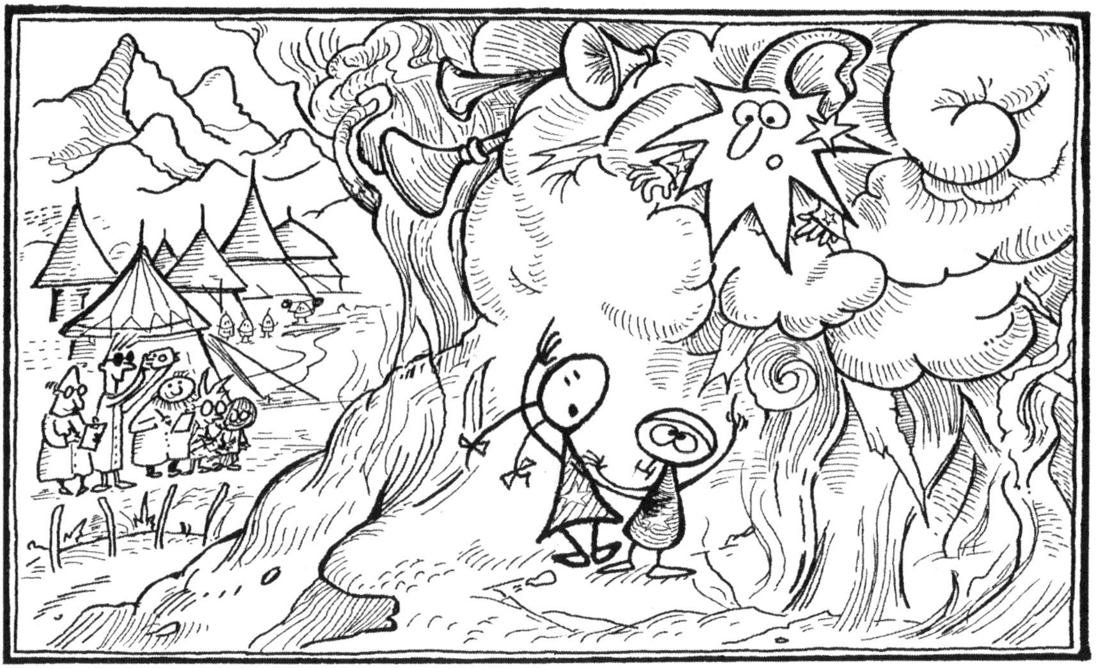

Published in 2011 by The Inamorata Press
18 Rosebery Road, Dursley, Gloucestershire GL11 4PT, UK

Text © Matthew Watkins 2011
Illustrations © Matt Tweed 2011

Matthew Watkins has asserted his right to be identified
as the author of this work in accordance with the
Copyright, Designs and Patents Act, 1988.

ISBN 978-0-9564879-1-9

All rights reserved. For permission to reproduce any part of this
book in any form whatsoever, please contact the publisher.

A CIP catalogue record for this book is available from the British Library.

Excerpts from "Prime Territory" by Enrico Bombieri reproduced with the permission of the New York Academy of Sciences. Where copyright permissions have not been explicitly granted to reproduce other short passages of text, all reasonable efforts have been made to communicate with copyright holders and secure these permissions.

The frontispiece is adapted from Hans Holbein the Younger's
16th Century woodcut *Moses on Mount Sinai*.

Printed and bound by Lulu (www.lulu.com).

www.secretsofcreation.com

Secrets of Creation
Volume Two
The Enigma of the Spiral Waves

words by
Matthew Watkins

pictures by
Matt Tweed

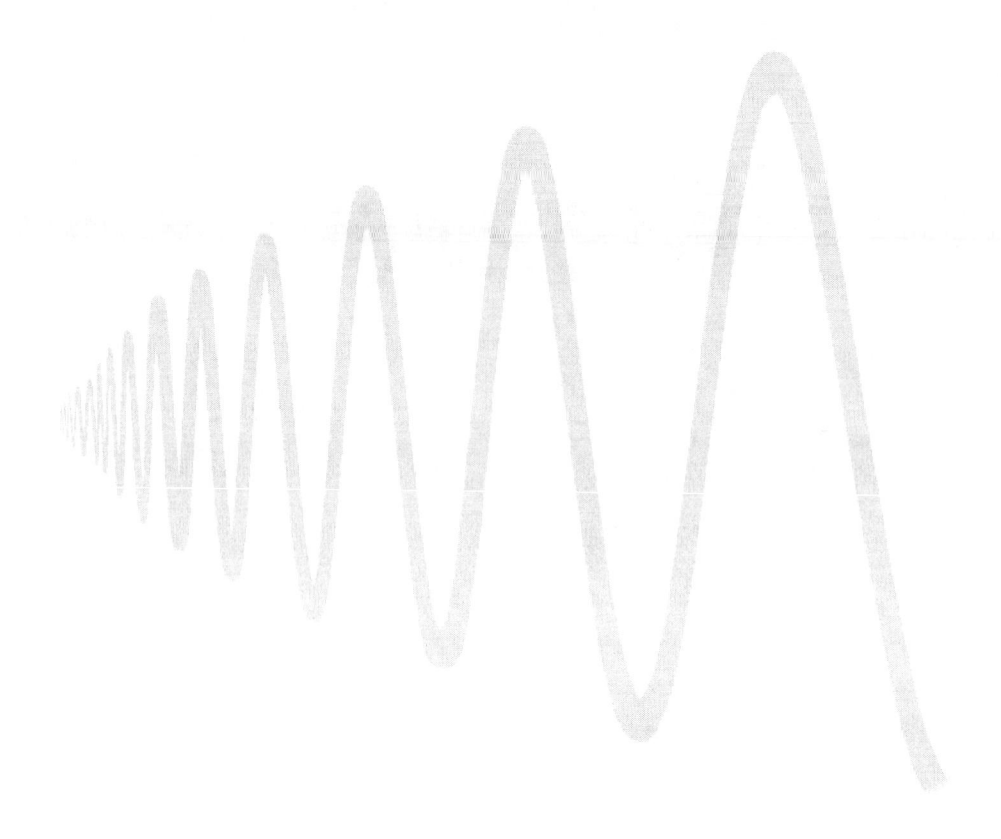

To me, that the distribution of prime numbers can be so accurately represented in a harmonic analysis... tells of an arcane music and a secret harmony composed by the prime numbers.

Enrico Bombieri, 1992

table of contents

a re-introduction	1
16. number systems within number systems *within number systems within number systems…*	5
17. The zeta function *Bernhard Riemann's big discovery*	29
18. back to prime numbers *remember them?*	43
19. the Riemann zeta zeros *a most mysterious splatter*	67
20. zeta zeros and spiral waves *and exactly how they're related*	81
21. the primes and the zeros *two sides of a coin (but what's the coin?)*	99
22. philosophical interlude *phew! put the kettle on…*	119
23. the Riemann Hypothesis *what everyone seems to want to know about*	139
24. reformulations of the Riemann Hypothesis *other ways of saying the same thing*	151
25. the significance of the Riemann Hypothesis *does this really matter and, if so, why?*	169
26. what's this *really* about? *do you really expect me to answer that question?*	189
27. a spectrum of vibrations *the Riemann zeros are vibrations…but of what?*	197
notes	212
appendices 10–14	226

For Elise, Laurie and Sophie.

The Enigma of the Spiral Waves is the second volume of the *Secrets of Creation* trilogy. It follows Volume 1: *The Mystery of the Prime Numbers*. The final volume, *Prime Numbers, Quantum Physics and a Journey to the Centre of Your Mind*, will eventually follow.

This volume is written with the assumption that you've recently finished reading Volume 1, or at least that the ideas which it contains are relatively fresh in your mind. If they aren't, then I'd highly recommend having a copy of the first volume available to consult as you read this one.

As in the first volume, I continue to avoid using mathematical notation in the main text, although quite a bit shows up in the appendices. These are there for readers interested in following up certain things in more detail. (All of the appendices in this volume require a certain amount of familiarity with mathematical notation and the reasoning behind its manipulation.)

The notes contain a number of website addresses. If you find any of these to have become inaccessible, the Internet Archive's "Wayback Machine" at www.archive.org is a useful tool for recovering old versions of webpages.

More information on the *Secrets of Creation* trilogy, further web links, additional resources and an ever-expanding list of thanks and acknowledgements can be found at www.secretsofcreation.com.

Matthew Watkins
Canterbury, 2011

a re-introduction

We first need to remember where we left off at the end of Volume 1. The main points which we'd explored were these:

✫ Underlying the system of counting numbers is a distribution of prime numbers.

✫ There's no obvious "pattern" in this distribution...

✫ ...but there *is* a (statistically) regular "thinning out".

✫ The actual distribution of primes deviates from this "average" thinning out.

✫ This deviation can be represented as a "messy waveform".

✫ This waveform can be broken down into a sum of infinitely many "spiral waves".

We were left, on page 293, confronted with a list of the (entirely random looking) frequencies of the first dozen or so of these "spiral waves", pondering where on earth they'd come from. This is the main question we'll be exploring in this volume, along with some further questions that this exploration naturally leads on to.

In a brief 1859 paper[1], Bernhard Riemann, in a dazzling burst of mathematical brilliance, supplied the necessary "big new idea" which was needed to prove the Prime Number Theorem (this having been around in the form of an unproven conjecture for over sixty years at that time). Eventually, Riemann's idea was developed to the point that proofs of the PNT, built on his work, were announced separately in 1896 by Jacques Hadamard and Charles de la Vallée Poussin.

Riemann's paper also brought to light a whole new range of problems and questions, one of which is now widely considered to be the most important unsolved problem in mathematics.

His "big new idea" was something now known as *Riemann's zeta function* – or, variously, "the Riemann zeta-function", "Riemann zeta", "zeta" or just "ζ" (the Greek letter *zeta* which Riemann used to denote it).

It's going to take a few more chapters before we can actually define this, but in the meantime, here are some (admittedly rather vague) things that can be said about it, to give you at least some idea of what kind of thing it is:

✭ It's intimately related to the distribution of prime numbers.

✭ It can be precisely defined in terms of the sequence of primes.

✭ Knowledge about the zeta function can be used to learn more about the primes and *vice versa*. Riemann used it to study what I called the "prime count deviation", something we first met in Chapter 12.

✭ It's going to provide some context for those mysterious frequencies we saw at the end of Volume 1.

✭ In contrast with the naive view of the primes as a collection of numbers to be viewed as isolated specimens, I've encouraged you to think in terms of the *distribution of primes*, which is a *single thing*. The zeta function (also a "single thing") in various ways encapsulates, encodes, characterises and mirrors the distribution of primes. It could perhaps be thought of as "the other side of the coin" (primes on one side, zeta on the other). The zeta function is something like the unseen "force" steering the school of "prime-fish" which we visualised back in Chapter 8. It somehow "orders" or "organises" the primes (although those words suggest something *taking place over time*, which could be confusing, so we must remain aware that language is being used loosely here, trying to convey something quite subtle).

✫ Whereas the idea of prime numbers is readily taught to, and understood by, young children, it normally requires several years of university-level maths before someone is ready to develop a basic understanding of the Riemann zeta function.

✫ What's widely seen as the most famous, most important *and* most difficult unsolved mathematical problem of the last century-and-a-half concerns the zeta function, so it's by no means fully understood (we'll get to that in Chapter 23).

✫ Mathematicians now study all kinds of other "zeta functions", most of which are not directly related to the study of prime numbers, although a great many of these can be linked mathematically to the original zeta function (Riemann's).

✫ The Riemann zeta function has been variously described as the most challenging, mysterious, infuriating, cruel, unusual and beautiful object[2] in all of mathematics.

As the name suggests, it's a *function*. We met some functions in Volume 1, but this one's a function of a somewhat different variety. So, we'll first need to make a diversion in order to broaden our concept of what a "function" can be because Riemann's zeta function inhabits a mathematical realm which most people are entirely unfamiliar with.

In fact, to understand this function and, through it, those mysterious frequencies associated with the spiral waves underlying the distribution of primes – *and thereby underlying the number system itself* – we'll have to go right back to basics.

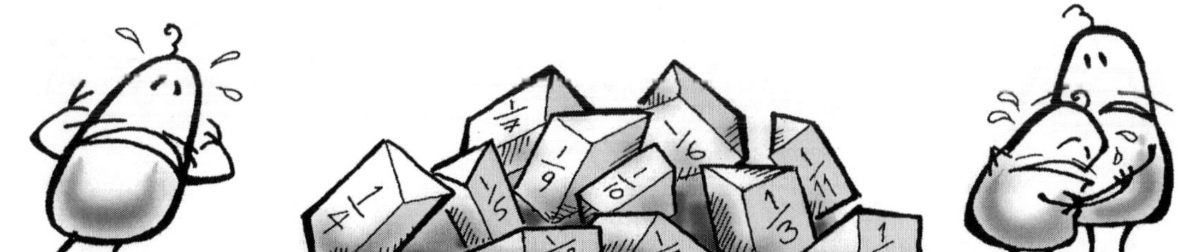
1/7
1/6
1/9
1/10
1/4
1/5
1/3
1/11
1/12
1/8
1/2

Chapter 16
Number Systems within Number Systems

Throughout the first volume, I often referred to *the* "number system", but there are in fact *many* number systems, as we shall now see.

I'm going to use a geometric approach to explain these number systems. Mathematicians would tend to explain them much more precisely but in a more abstract and harder-to-imagine way. Nonetheless, they would concede that the images I'll present here are, in essence, describing the same thing.

What we're about to see is just a representation – numbers *are not* points on a line – but, if properly understood as a representation, I've found that this is the easiest way for most people to grasp the ideas.

We start with an unmarked line:

Pick any point. We'll call that "0":

Now pick any *other* point. We'll call that "1":

This "1" point could have be chosen to be on the left of "0" but people tend to think of it as being to the right. It doesn't actually matter, as we could just turn the page upside down – a line has no built-in left or right, it depends which way you look at it.

Everything now follows. The distance between 0 and 1 will act as our "unit". Notice how it doesn't matter what the unit is – it's completely arbitrary, because before you choose the second point (the "1"), *there's no way of measuring anything*.

All choices of "1" (remember that it can be any point other than the point chosen for "0") are basically the same in the end. Just imagine turning your page around if necessary, moving it closer to you or farther away. In that way, "1" can always be seen to lie to the right of "0" and at any distance from it you want.

We now "grid out" the line to the right by using our unit to mark off an endless sequence of points which we'll call "2", "3", "4", "5", "6", "7",... So we now have 0 and the counting numbers. Mathematicians denote the collection of all counting numbers "ℕ" (they usually call them *natural numbers*). Confusingly, ℕ is sometimes intended to include 0 and sometimes not. For us, it's *not* intended to include 0.

We next "grid out" the line to the left. These points mirror 1, 2, 3, 4, 5,... and are usually denoted "−1", "−2", "−3", "−4", "−5",... These are *negative numbers*, which showed up briefly in Chapter 3. In fact, all the points to the left of zero correspond to negative numbers. −1, −2, −3, −4, −5,... are specifically called negative *integers*. The numbers most people are familiar with are *positive numbers* and correspond to points to the right of 0. Each of these has a "mirror image" to the left of zero, a negative number (written the same as the original positive number, but with a minus sign):

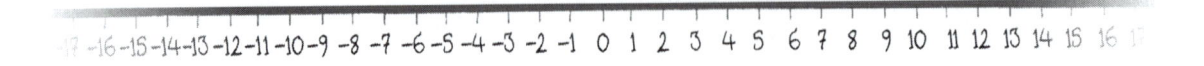

−17 −16 −15 −14 −13 −12 −11 −10 −9 −8 −7 −6 −5 −4 −3 −2 −1 0 1 2 3 4 5 6 7 8 9 10 11 12 13 14 15 16 17

The set of "grid points" we now have (zero, the counting numbers 1, 2, 3, 4, 5,...

and the negative integers −1, −2, −3, −4, −5, ...) collectively make up the *integers*. Pairs of integers can be added and multiplied according to certain rules, resulting in further integers, so we have a number system. Mathematicians denote it by "ℤ".

Notice that ℤ *includes* ℕ – all counting numbers are integers (but not *vice versa*).

We'll now restrict our attention to the piece of number line between 0 and 1, the length of which is acting as our unit.

Divide it into two pieces by marking a point midway between 0 and 1. That point, of course, represents the number 1/2 (also written as "0.5").

Now divide it into three pieces by marking a pair of points in the appropriate locations between 0 and 1. These points correspond to 1/3 and 2/3 (also written as "0.333..." and "0.666...").

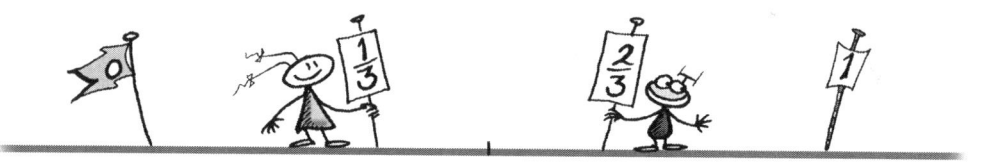

Next divide it into four pieces. You'll find that one of the three points you need to mark has already been marked (that's 1/2). The others are 1/4 and 3/4 (also written as "0.25" and "0.75"). You can see how 1/2 is the same as 2/4 (or "two quarters").

We then divide it into five pieces...

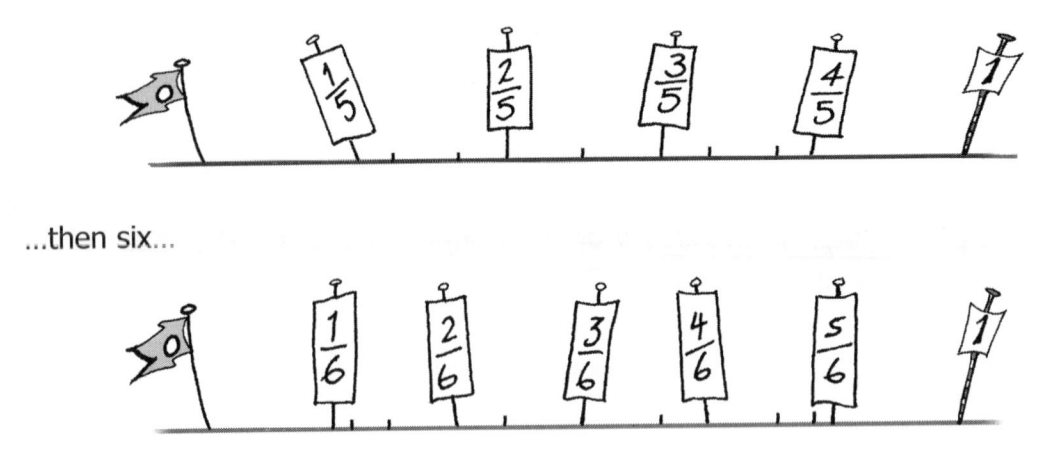

...then six...

...and so on.

Imagine doing this for *all* the counting numbers and in this way gradually filling up the space between 0 and 1 with marks. It would gradually become a fuzzy grey, eventually approaching black, indistinguishable from a solid line segment. We'd then have marked *all possible fractions between 0 and 1*. Now imagine picking up this piece of number line between 0 and 1 (full of marked points) and repeatedly "printing" it along the "grid", that is, between 1 and 2, between 2 and 3, between 3 and 4, *etc.*

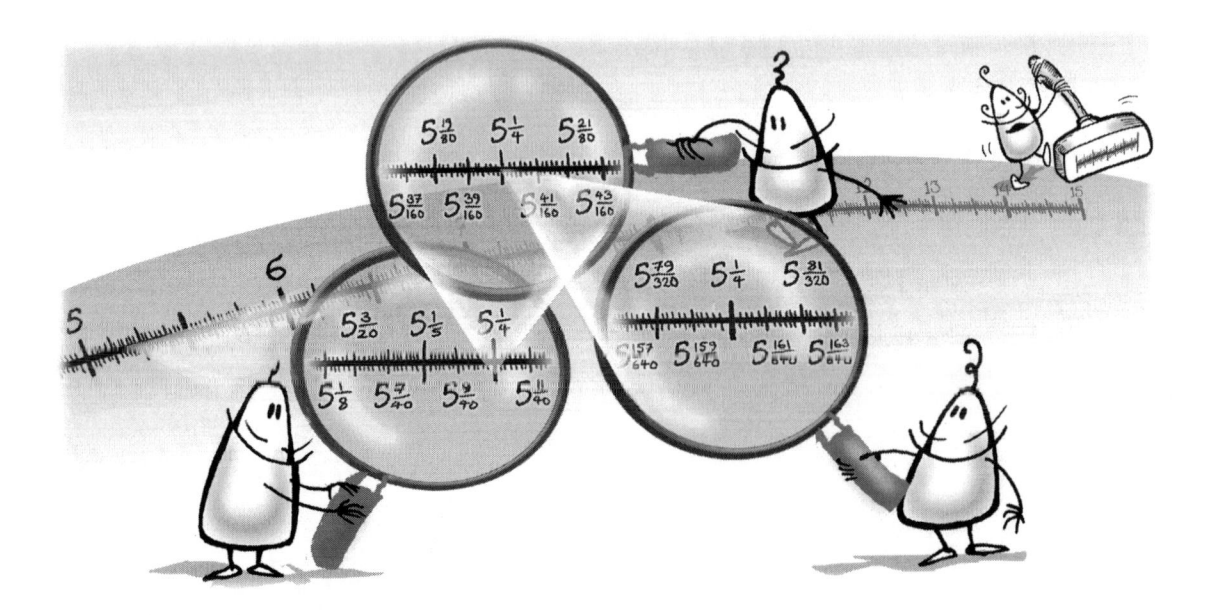

8

Of course, we'd also have to print in the negative direction. Alternatively, we could imagine "tiling" the number line with this particular patterned "tile".

The result of this would be *all* possible fractions (any integer divided by any counting number), so all of the following examples would be included:

12	which is 12 divided by 1
$\frac{1}{2}$	which is 1 divided by 2
$\frac{25}{4}$	which is 25 divided by 4
$-\frac{17}{5}$	which is −17 divided by 5
$\frac{21986}{17}$	which is 21986 divided by 17
$-\frac{4}{9}$	which is −4 divided by 9
$-16\frac{29}{55}$	which is −909 divided by 55
-7	which is −7 divided by 1
0	which is 0 divided by 1, or by any other counting number
$\frac{1}{10007}$	which is 1 divided by 10007

Pairs of these can be added and multiplied according to clearly defined rules, producing further fractions, so we now have another number system. It's denoted "ℚ" and its numbers are called *rational numbers*. Notice that ℚ includes ℤ (which itself includes ℕ). Every integer is a rational number, but not every rational number is an integer.

You might think that ℚ would cover the whole line, that every point on the line must correspond to a number in ℚ. But, perhaps surprisingly, *this isn't true*. There are holes – a *lot* of holes, even if you can't see them[1].

Here we see a few easily accessible examples of "holes in ℚ":

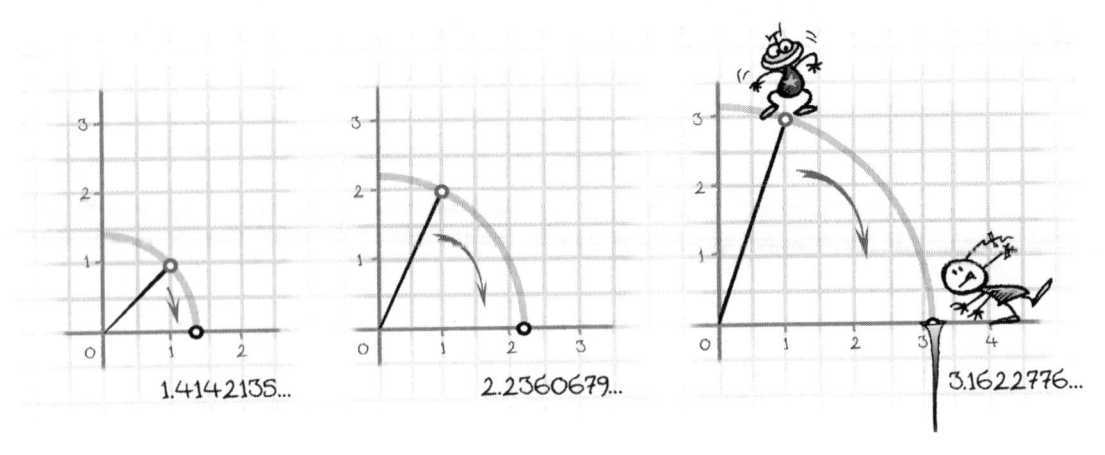

1.4142135...   2.2360679...   3.1622776...

According to the famous theorem of Pythagoras, the diagonal of a square with side length 1 equals the square root of 2 (the number which when multiplied by itself gives 2). Similarly, the diagonal of a rectangle with sides 1 and 2 equals the square root of 5, and that of a rectangle with sides 1 and 3 equals the square root of 10. All of these can be shown not to be expressible as fractions.

There are infinitely many "holes" in ℚ, but they're not generally describable in this visual way – these are just some examples which are fairly easy to "see". Another one can be located by stretching out half of a circle along the number line:

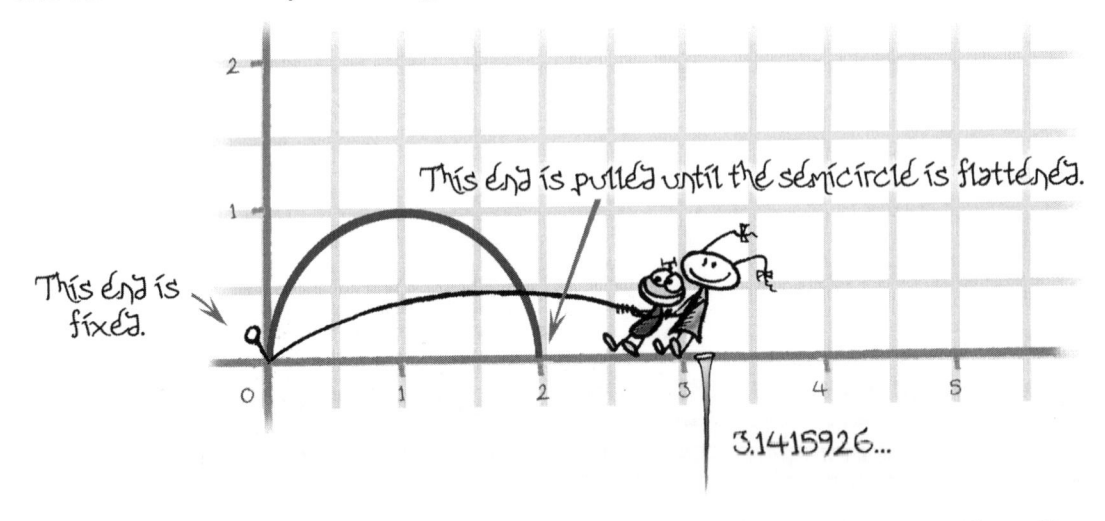

This end is fixed.

This end is pulled until the semicircle is flattened.

3.1415926...

The endpoint on the right will be π (that's 3.14159...). This lies between the rational numbers 3141/1000 and 3142/1000 and it's well known to be a hole in ℚ. It's a hole because it's *not* possible to express π as a fraction (this was proved in 1761[2]).

10

Another famous hole in $\mathbb{Q}$ is the number "*e*" which we met in Chapter 9. Recall that this is about 2.718 (so it lies between the rational numbers 2718/1000 and 2719/1000). The fact that it's impossible to express *e* as a fraction was proved in 1737[3].

These numbers which don't belong to $\mathbb{Q}$ but which still seem to have some sort of existence (in as much as any of these types of numbers can be said to "exist") are called *irrational numbers*. There are infinitely many of these, just as there are infinitely many rational numbers[4], but they *don't* constitute a number system[5].

As you might be suspecting, there's yet another number system on the line, one which includes $\mathbb{Q}$. In fact, there are *infinitely many* such number systems which include $\mathbb{Q}$, but until the 1890s, only one was known about. It's usually denoted "$\mathbb{R}$" and is routinely studied by mathematics students at university level. It includes $\mathbb{Q}$ together with all of the irrational numbers I've described. We'll soon find out what the "R" stands for.

The number system $\mathbb{R}$ is what's called a *completion* of the number system $\mathbb{Q}$. The idea of a "completion" is quite subtle, but mathematicians have a precise definition that they work with[6]. Very roughly, we can think of a completion of $\mathbb{Q}$ as a "meaningful way to fill in the holes". Before the 1890s, there was only one known way to "complete" $\mathbb{Q}$. Then, it was discovered that there are *infinitely many other ways of completing it, each producing a different number system*. The (really quite strange) results of these other completions are called *p-adic number systems*, where "p" stands for "prime". For each prime number, there's one of these number systems: a 2-adic number system, a 3-adic number system, a 5-adic, 7-adic, 11-adic,... number system. Each one uses its particular prime number as part of a process to "fill the holes" in $\mathbb{Q}$[7], and in each case we get an entirely different number system (which includes $\mathbb{Q}$). We won't be too concerned with these *p*-adic number systems (they'll get another mention, but you won't need to know anything more about them). For now, we'll stick with the much more widely known, and seemingly "sensible", number system $\mathbb{R}$.

So, every point on the line corresponds to a number in the system ℝ and every number in ℝ corresponds to a point on the line. ℝ includes zero, positive numbers, negative numbers, integers, rational and irrational numbers. And, of course, it contains the prime numbers.

Pairs of numbers in ℝ can be added and subtracted, multiplied and divided according to clearly definable rules, resulting in numbers in ℝ. We'll now have a quick, informal look at how this works geometrically, as this will be helpful later on.

Adding and subtracting can be visualised in terms of walking certain distances on the number line like this:

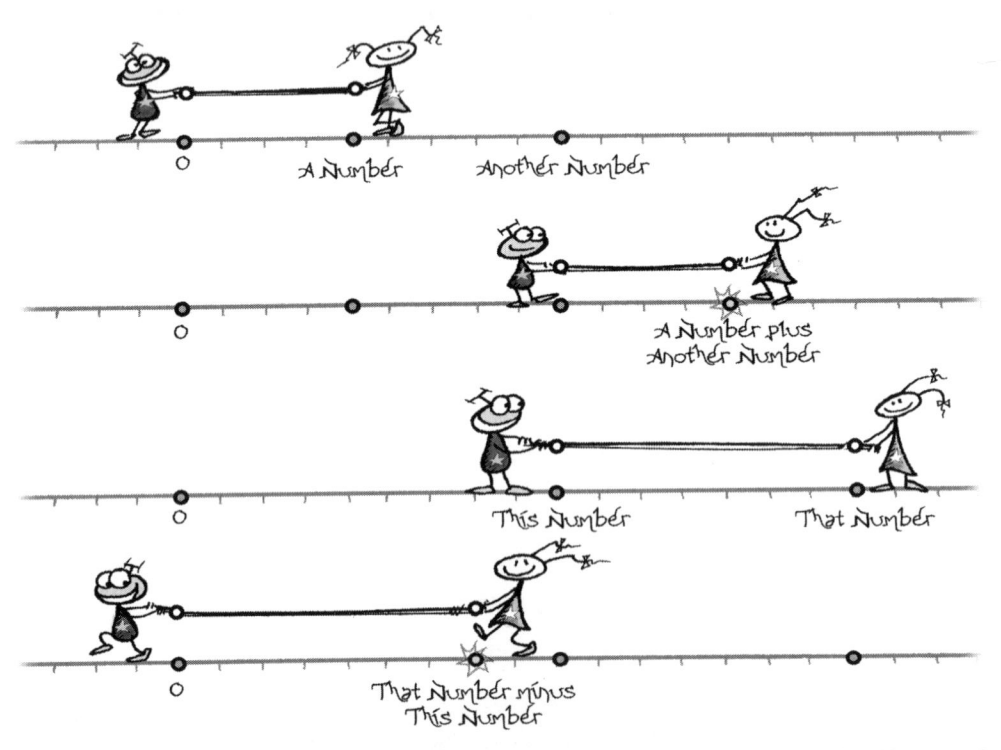

Addition: when zero becomes "Another Number", what does "A Number" become? Subtraction: when "This Number" becomes zero, what does "That Number" become? This approach works for *any* pair of numbers, positive or negative.

Multiplying and dividing, seen geometrically, are all about magnification (or "dilation"). This can most easily be understood by visualising an elastic rope pegged to 0.

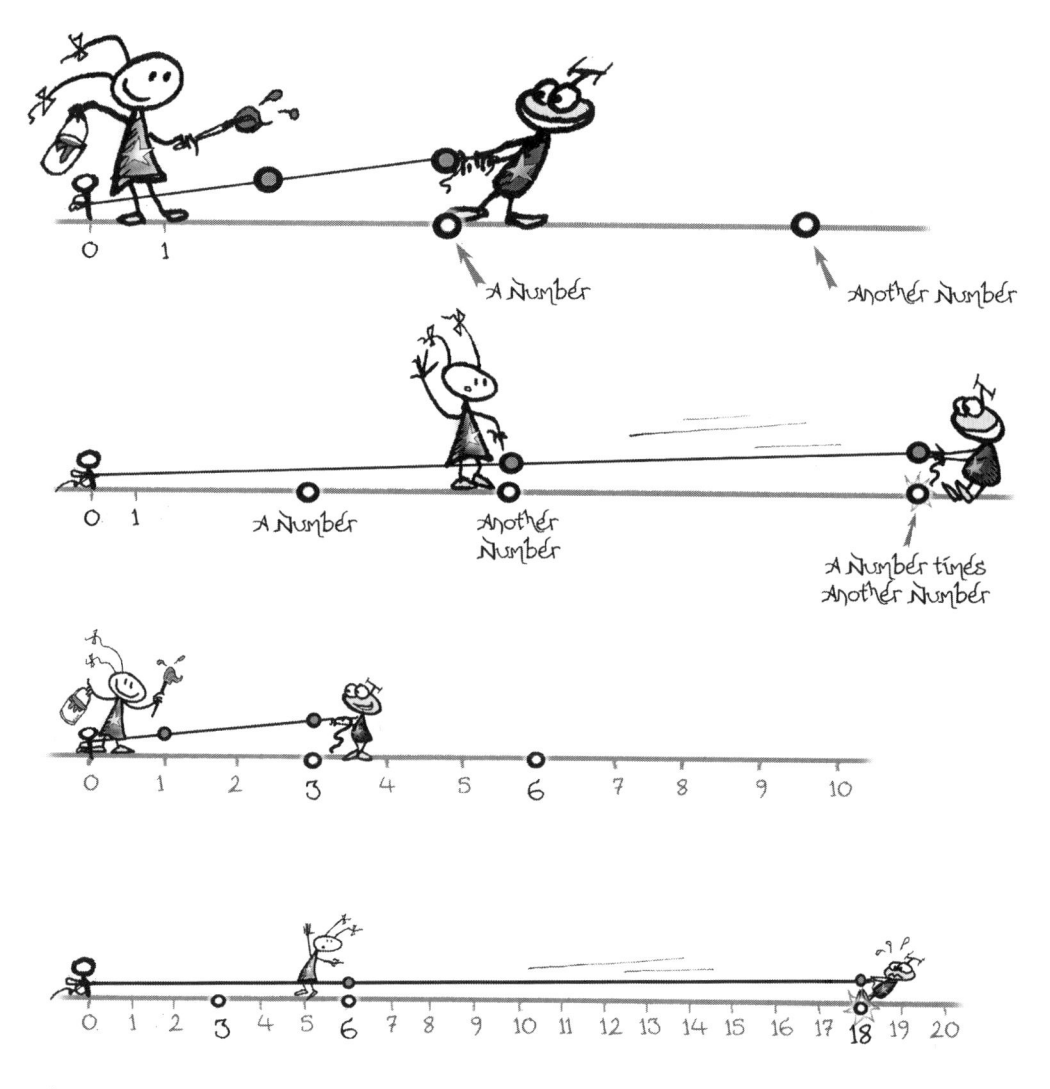

When confronted with "6 × 3", you can think of it as posing the following question: "if 1 is stretched so that it becomes 6, what does 3 become?"

It's worth observing that with adding and multiplying, it doesn't matter which way around we do it (this was mentioned in the last volume – they're "commutative operations", so, for example, 5 + 7 = 7 + 5 and 12 × 19 = 19 × 12). This doesn't work for subtraction, though.

Division doesn't have this property either. If you follow the next visualisation carefully, you might be able to see why.

13

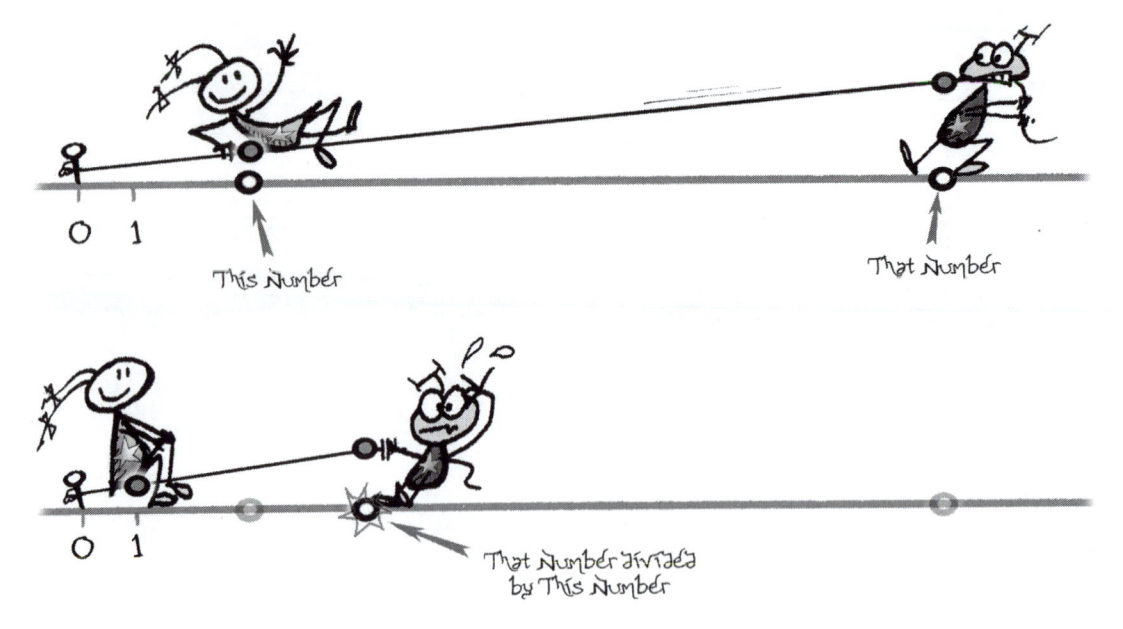

Division: When "This Number" shrinks/stretches to 1, what does "That Number" shrink/stretch to?

Suppose we now employ a bit of lateral thinking (literally) and, choosing a point *on* the page but *off* the line...

...we ask "what number does this represent?" If points on the line represent numbers, then what about other points on the page which aren't on the line? This might seem like an odd question. A straightforward answer would be "it *doesn't* represent a number — we've stepped outside the system of points-on-a-line-representing-numbers".

Surely this point cannot represent a number, as all possible numbers, positive and negative, whole and fractional, rational and irrational, are represented on the line.

Well, it turns out that there *are* numbers which these points represent. It was discovered in the 1790s that the number line is naturally "embedded" in a number *plane*. It's called the *complex plane*, and points in it correspond to what are known as *complex numbers*. That's "complex" not in the commonly used sense of "complicated" but in the original sense of the word, that is, "made of parts" – in this case, two parts. Each complex number, represented by a unique point in the plane, can be thought of as being built out of *a pair of numbers from* $\mathbb{R}$, the number system represented by the line – the first number in the pair describes the point's "east-west position" in the plane, the other describes its "north-south position".

This "number plane" may seem like a bit of an unnecessarily weird or complicated idea, but it's perfectly possible to describe such a thing in a way which makes sense mathematically. Its points correspond to a wider class of "numbers" which includes the usual numbers from $\mathbb{R}$, but also many other "complex numbers" which aren't in $\mathbb{R}$ (and which can't be thought of as "amounts" of something). Even though this extended concept of numbers may not fit with your ideas about what "numbers" should be like, the system makes mathematical sense because you can add, subtract, multiply and divide complex numbers according to clear rules, just as you would the usual ones, and you end up with a kind of "arithmetic of complex numbers" which is *consistent* – that is, it won't ever contradict itself.

Yet again, we've expanded our number system. This system of complex numbers is denoted "$\mathbb{C}$" and it includes $\mathbb{R}$, just as $\mathbb{R}$ includes $\mathbb{Q}$, $\mathbb{Q}$ includes $\mathbb{Z}$, and $\mathbb{Z}$ includes $\mathbb{N}$. Numbers in $\mathbb{R}$ can be thought of as *special examples* of complex numbers.

Complex numbers can be thought of as "generalised" or "two-dimensional" numbers (in the sense that the numbers we're used to, the numbers in $\mathbb{R}$, can all be understood as points on a number line, and a line is one-dimensional). The two dimensions correspond to the two parts of a $\mathbb{C}$ number, the pair of $\mathbb{R}$ numbers which play the role of its "coordinates", these providing a systematic way to describe the location of a point in a plane. Here's how. Our plane already contains a line representing $\mathbb{R}$. We draw another line, at a right angle, through 0, dividing the plane into four regions.

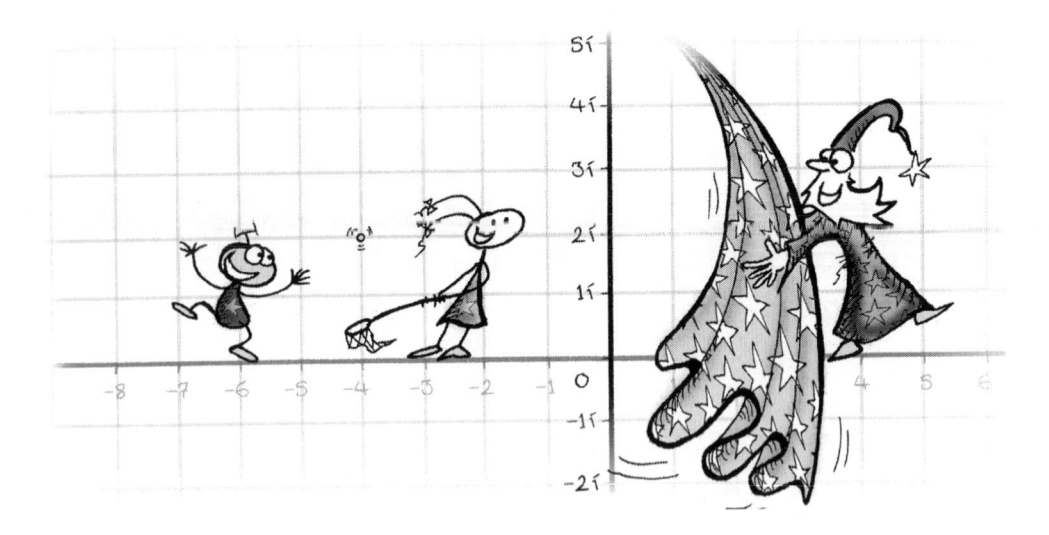

We also think of this as a number line, but we always write "*i*" after these numbers, as shown. Any point on this plane can now be systematically described using two numbers: one from the familiar horizontal number line, which tells us "how far left or right" the point is, and another from the new vertical number line, which tells us "how far up or down" the point is.

Here are some examples:

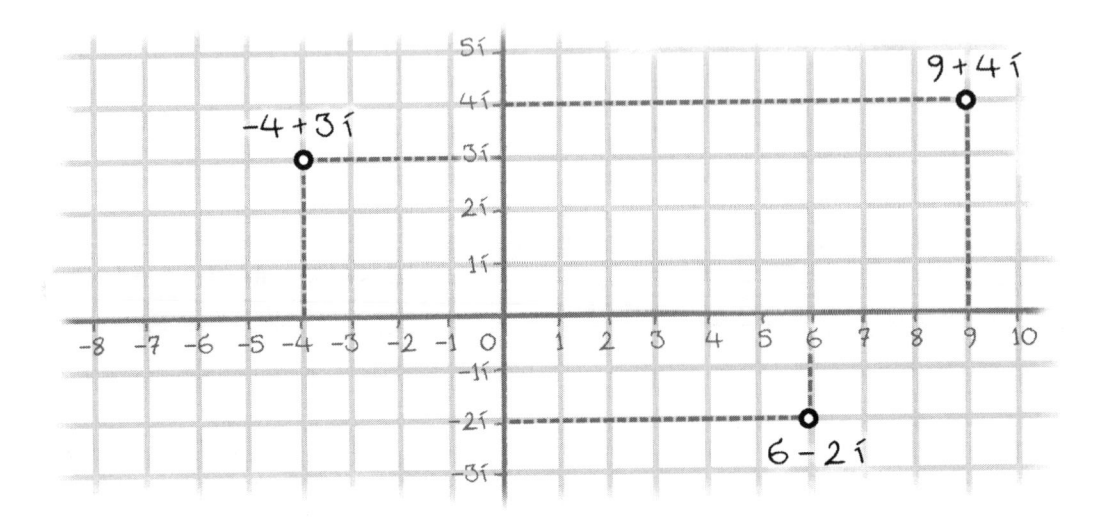

16

You might be wondering what the "i" stands for. And you might still be wondering what the "$\mathbb{R}$" stands for. The answer confuses some people, so first we'll need a little bit of historical background. When these numbers "off the line" were first discovered (some people might say "invented" – this is a subtle point), the general feeling about them was that "they're not *real* numbers, they're just a product of our imagination". The question of in what sense numbers *of any kind* are "real" is a very difficult matter which philosophers continue to debate. But, in any case, as a result of this initial reaction, numbers "on the horizontal line" have become known as *real numbers* (hence "$\mathbb{R}$") and numbers "on the vertical line" are known as *imaginary numbers* (hence the "i").

In the early 21st century, the general feeling in the mathematical community is that both types of numbers are equally real (or equally imaginary, depending on how you look at it). But the names have stuck. So we'll be talking about "real" and "imaginary" numbers, but we could just as well be talking about "horizontal" and "vertical", "type A" and "type B", or "green" and "purple" numbers.

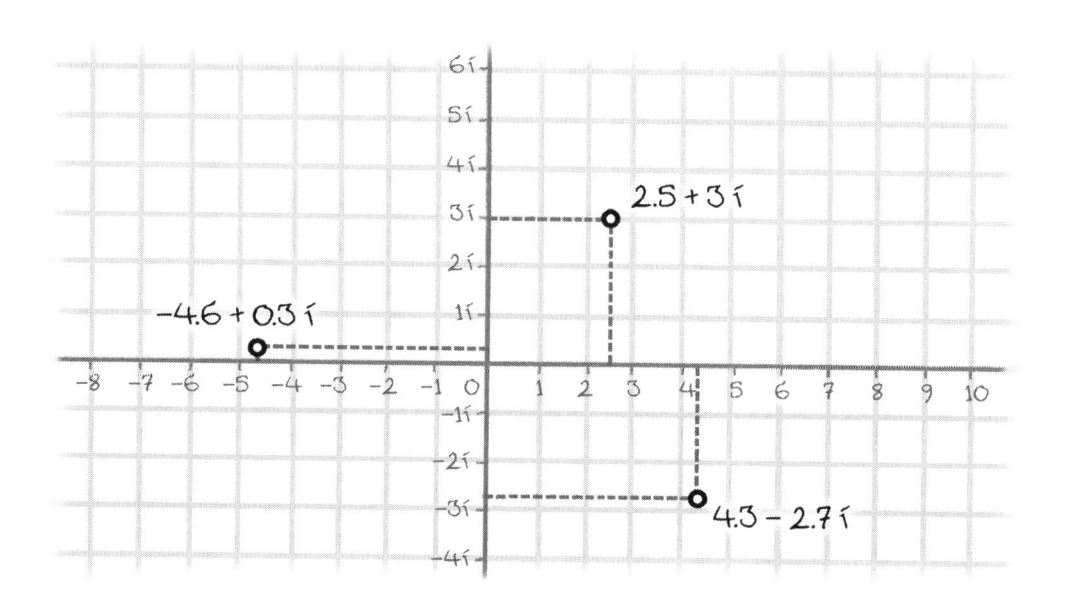

In the previous examples, real and imaginary parts are all integers. Here they're all rational numbers. But these numbers could equally well be irrational.

A number represented by a point *on* the horizontal axis is a complex number whose "imaginary part" is zero — it's a special example of a complex number, because it's also a real number. A number represented by a point on the *vertical* axis is a complex number whose "real part" is zero. These are also "special examples" of complex numbers, sometimes called "pure imaginary" numbers. In general, a complex number won't be of either of these two special types.

It was the pure imaginary numbers which were the first to be considered historically, as a way of "forcing" solutions to simple equations like "$x^2 + 4 = 0$" which were previously seen as simply *not having* solutions.

How do we add, subtract, multiply and divide these strange new "complex" numbers?

We'll refer back to our earlier geometric descriptions of how to do these things in $\mathbb{R}$ and then attempt to extend the relevant procedures to $\mathbb{C}$.

Adding and subtracting are quite straightforward. Again, given two complex numbers, we have two points on the plane:

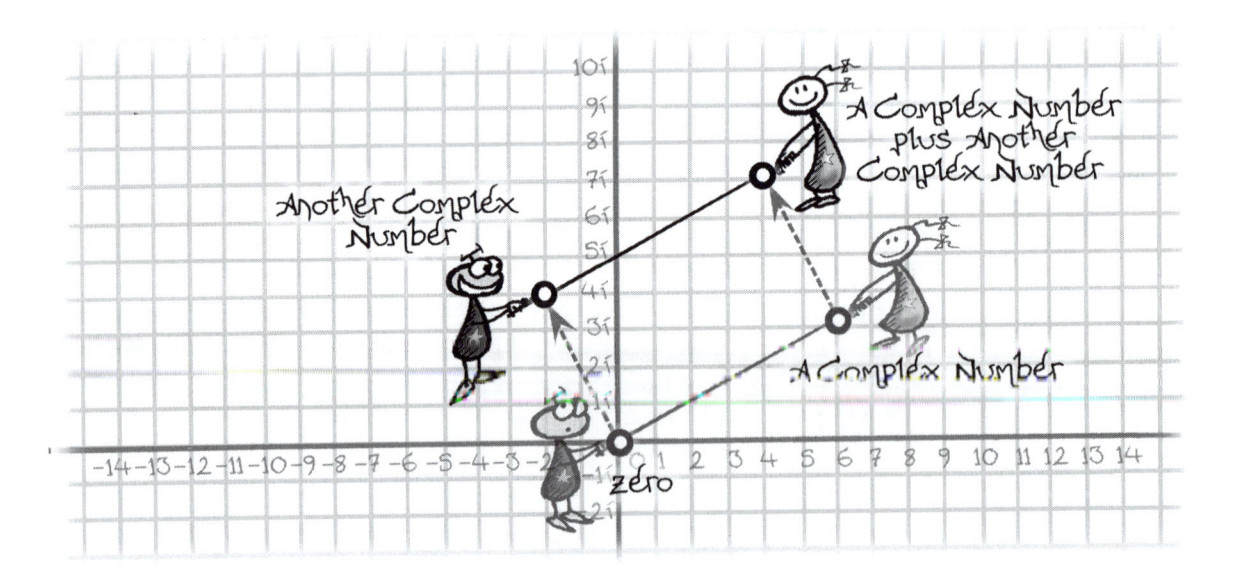

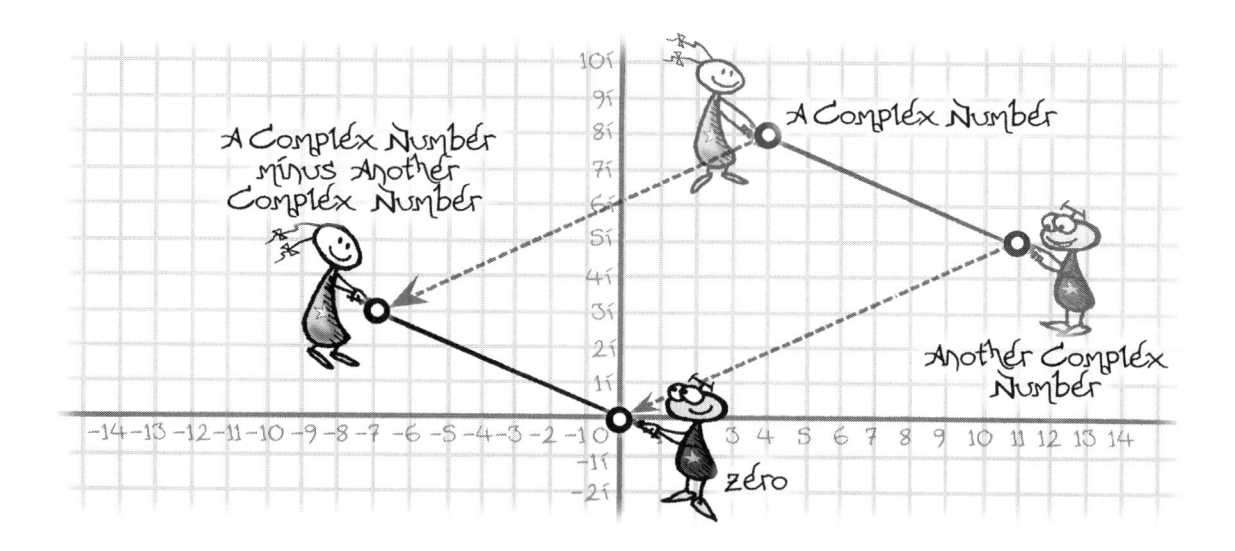

In the case of addition (opposite page), we slide from zero to one of the numbers and (keeping it aligned with zero) see where the other ends up. In the case of subtraction (above), we slide from one of the numbers to zero and (keeping it aligned with zero) see where the other ends up.

Notice that if both our given numbers happen to be real, then the procedure shown above for addition is exactly the same as the one shown earlier for ℝ. This new procedure would be said to *generalise* the original one. It *includes* the original one, but can be applied in a much more "general" way.

Also, if you look carefully at what's happening, you'll see that adding two complex numbers can be achieved by adding their real parts and adding their imaginary parts separately. The same kind of thing also works for subtraction. Multiplication, however, is not so simple.

On ℝ, we saw multiplication and division illustrated by procedures involving an elastic rope. We'll now generalise this procedure to ℂ.

The multiplication procedure for ℝ involved a kind of magnification or dilation. The multiplication procedure for ℂ will too, but it will also involve *rotation* (something which you can do when you're in a plane, but not when you're stuck on a line).

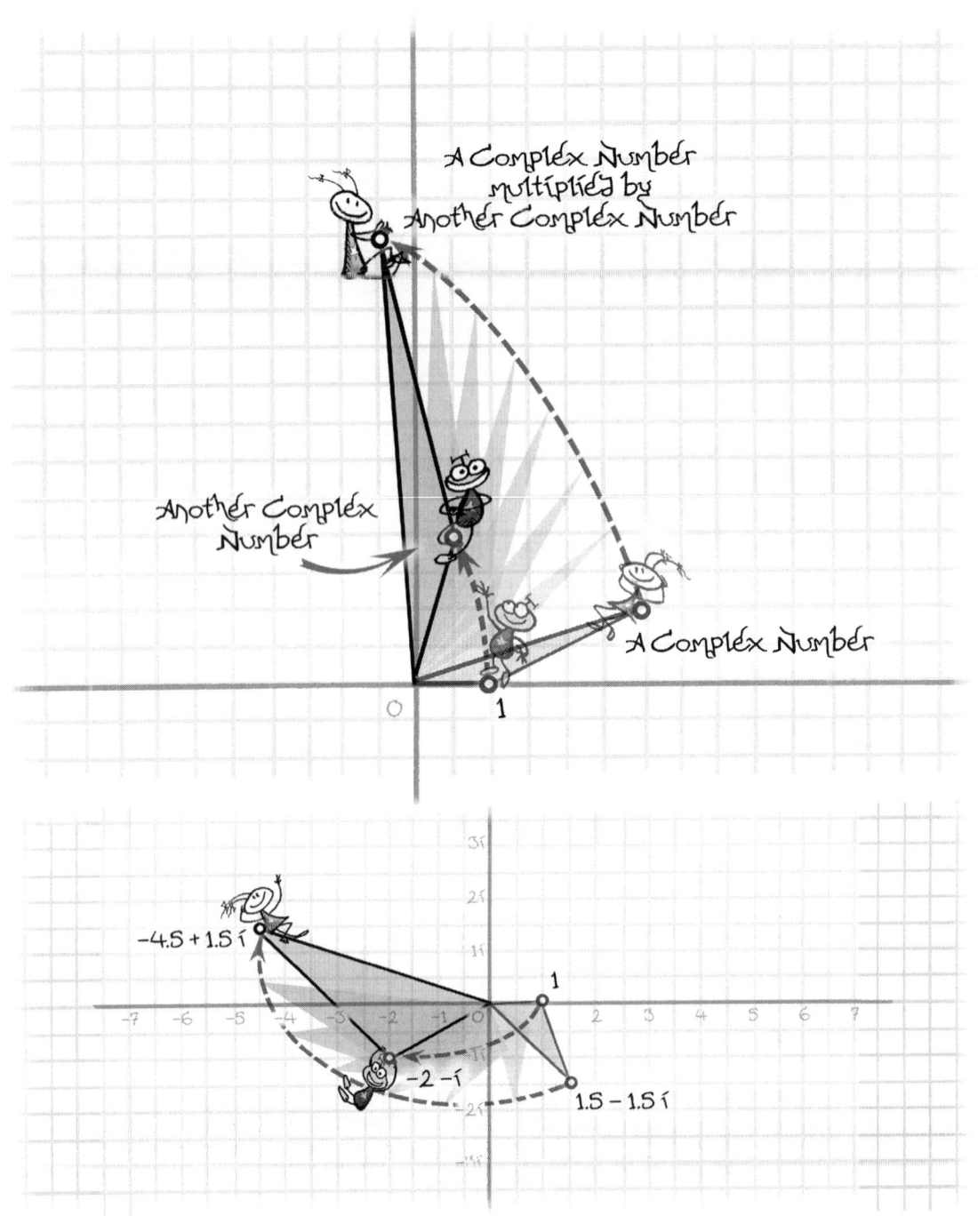

Multiplication: When 1 is stretched and rotated to "Another Complex Number", where does "A Complex Number" end up? Example: $1.5-1.5i$ times $-2-i$ equals $-4.5+1.5i$.

A few observations might help at this point.

One is that it doesn't matter in which order you multiply – if you reverse the roles of the two points in the visualisation we've just seen, you end up with the same answer.

Another is that if both of the complex numbers happen to be in $\mathbb{R}$ (that is, their points happen to lie on the horizontal number line), then this procedure reduces to the one we saw earlier for multiplication in $\mathbb{R}$. In $\mathbb{R}$, no rotation is necessary[8], the relevant triangle being completely "flat", but the procedure still works. If you have trouble seeing this, first try to imagine the procedure we just saw, but for two points which are *very slightly above* the horizontal line, and to the right of zero. Then imagine doing it for pairs of points which are closer and closer to the line (so that the triangle becomes flatter and flatter).

Finally, when multiplying two complex numbers, we end up with a third, and its distance from 0 can be found by multiplying the distances of the other two from 0. If we then draw lines from each of the three points to 0, we see that each makes a particular angle with the horizontal number line. The angle of the third can be found by *adding* the other two angles. The multiplication procedure illustrated opposite with the elastic triangle, *etc.* is just a geometric way of doing this.

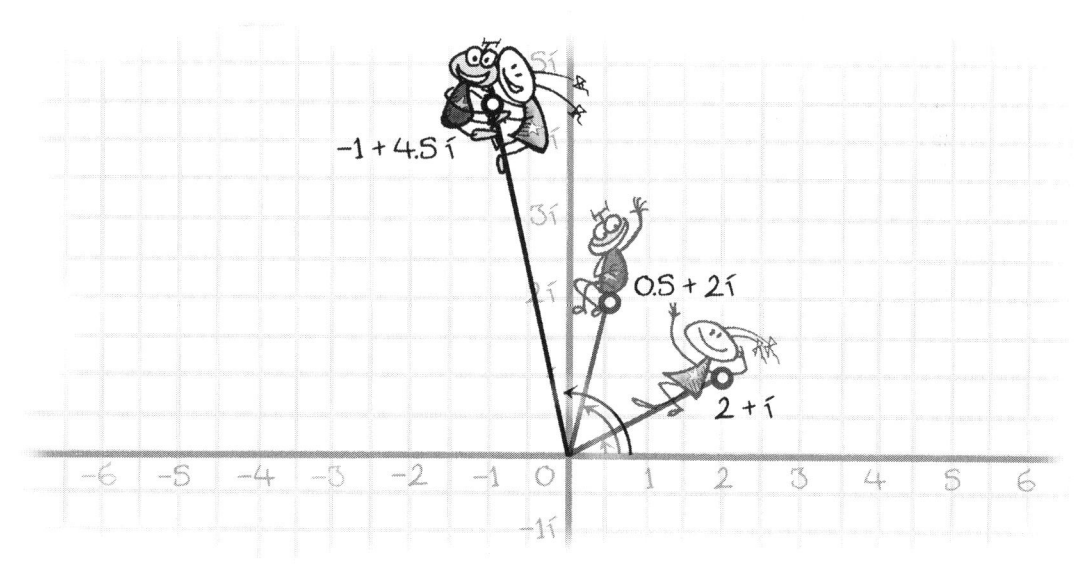

The distances here are approximately 2.236, 2.061 and 4.609 (first × second = third).
The angles are approximately 26.56, 75.96 and 102.53 degrees (first + second = third).

21

By far the most famous of the complex numbers is represented by the point which lies one unit up the vertical axis from 0. I've shown it as "1i" on the diagrams we've seen so far, but it's usually just called "i", also known as the *imaginary unit*. This was the first complex number that anyone ever considered. It's of particular interest because of what happens when you multiply it by itself.

According to the procedure we just saw, we start with this triangle...

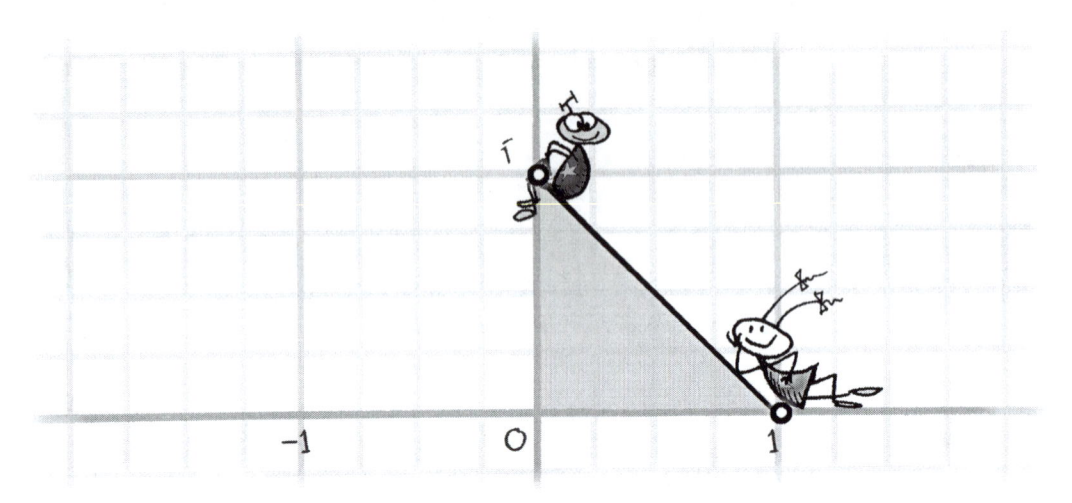

...which gets rotated like this (no stretching necessary for 1 to get to i):

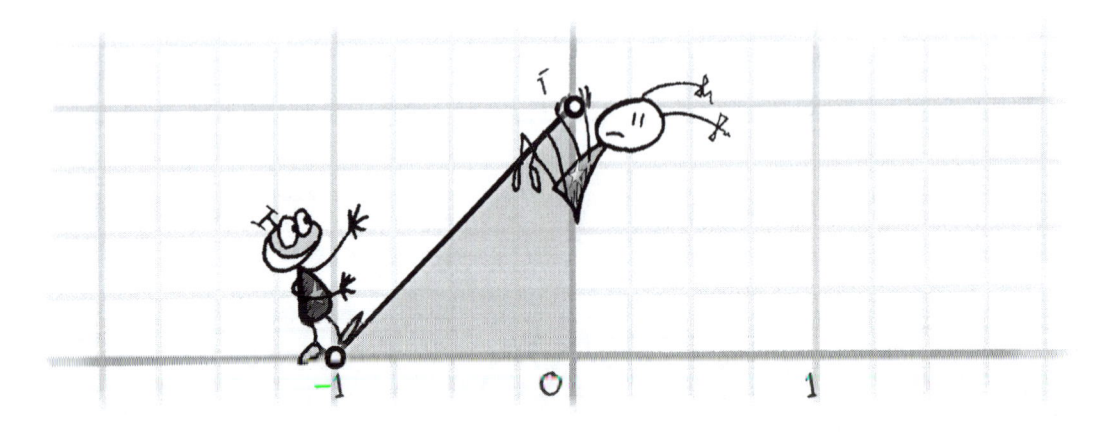

In this way, we see that

$$i \times i = -1.$$

i has been rotated a quarter turn anticlockwise around 0. The effect of multiplying *any* complex number by *i* is to rotate it a quarter turn anticlockwise around 0. Why is this? Well, I've explained that to multiply complex numbers, you can multiply distances from 0 and add angles. The point *i* has distance 1 from 0, so multiplying by that has no effect on the other distance (anything multiplied by 1 is left unchanged). And the point *i* is at a right angle to the horizontal ($\mathbb{R}$) axis, so adding that right angle to any other angle has the effect of a quarter turn anticlockwise:

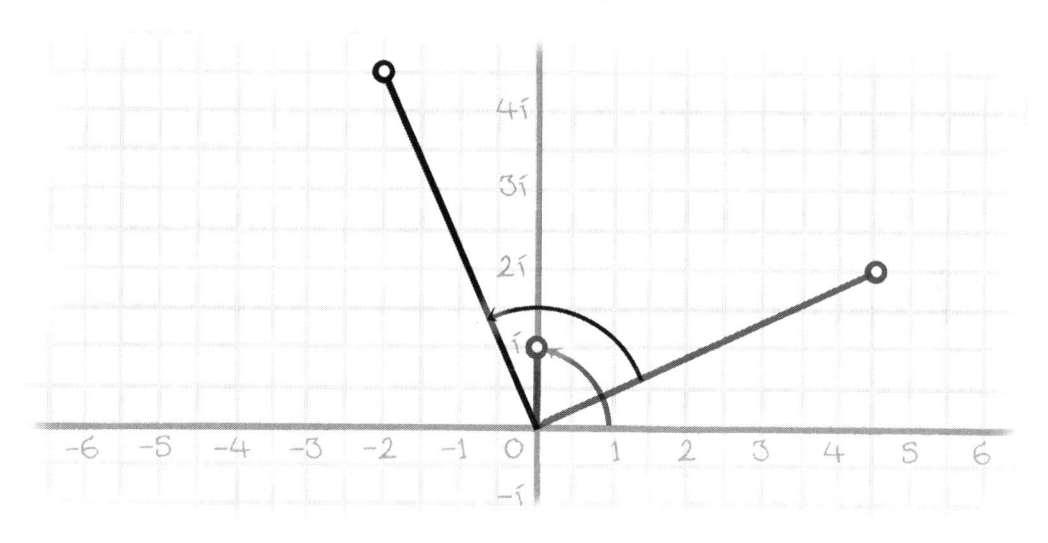

Here the relevant angles are approximately 23.96 and 113.96 degrees (note 23.96 + 90 = 113.96).

It's impossible to multiply any *real* number by itself to get −1.

Surprisingly, for some people,

$$-1 \times -1 = 1.$$

In fact, any pair of negative numbers multiplied together gives a positive number. This is often a source of confusion. It's possible to explain it in terms of debts (which can be thought of as negative amounts of money): having a certain number of equal debts simultaneously cancelled results in "gaining" a negative number of negative amounts of money which you can see is going to be a net positive outcome[9]. But we'll stick with our elastic rope as a more physical/visual approach to illustrating this.

What we saw earlier only works for two positive numbers, but the same visualisation can be easily adapted to the multiplication of any combination of positive or negative numbers. First, let's try one of each, say 3 × −7 (which equals −21):

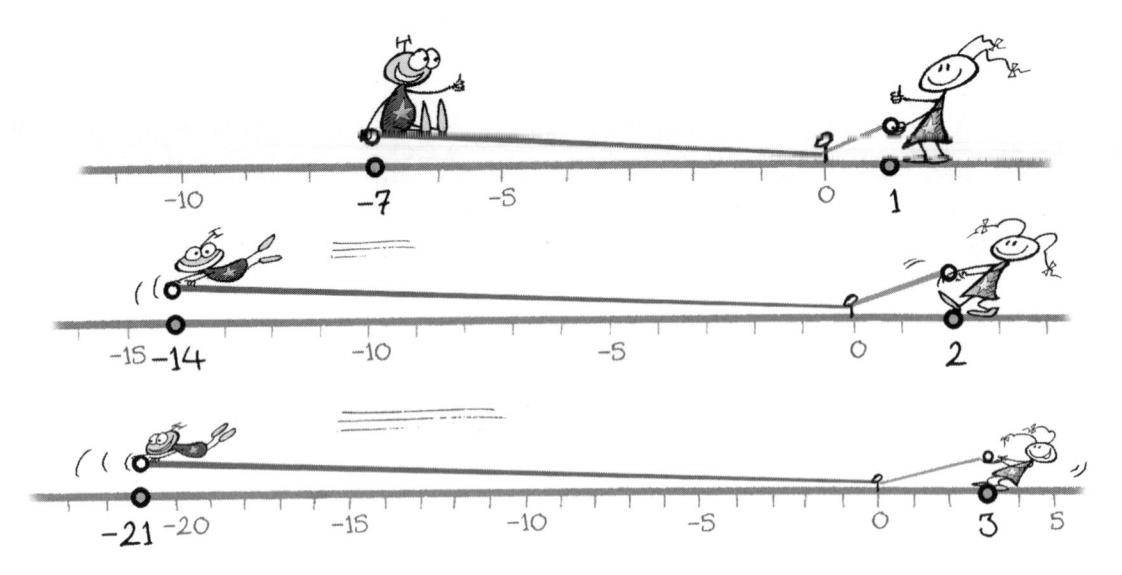

Physical reality doesn't work like this, but in these visualisations, the two sections of elastic rope are magically related, so stretching one section causes the other to stretch by the same magnitude.

Now let's try −3 × −7. The idea is to stretch the rope so that 1 ends up at −3. But these points are on opposite sides of zero. So, the obvious thing to do is to switch the ropes around, pull the original rope to the left, the "mirror image" rope then stretching out to the right.

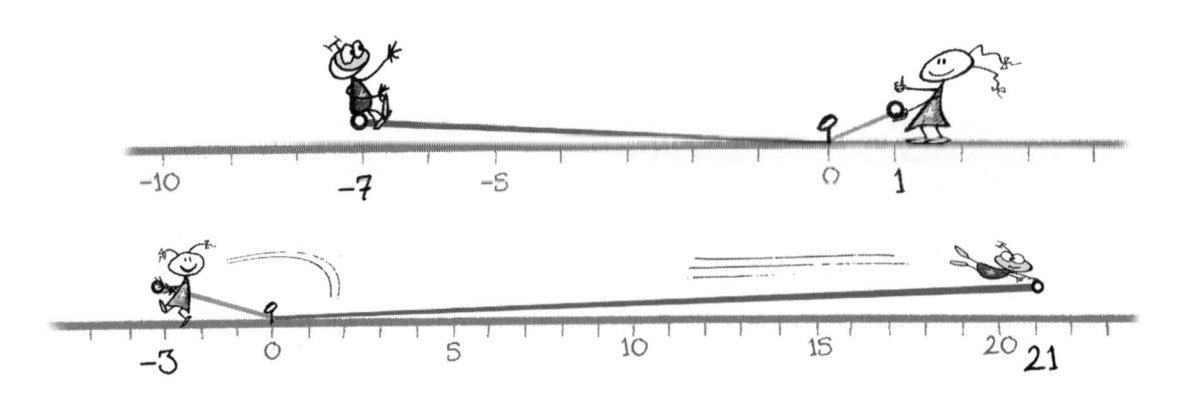

24

So 1 ends up at −3, and −7, having been flipped over to position 7 on ℝ, gets stretched to 21. So −3 × −7 = 21.

If you're struggling with this explanation, don't worry too much. We won't really need to *use* this. You'll just have to trust that there are very good mathematical reasons why:

positive × positive = positive

positive × negative = negative

negative × positive = negative

negative × negative = positive

Mathematicians trying to solve a certain type of "impossible" equation centuries ago realised that they *could* find solutions, but only if they had access to a number which, when multiplied by itself, gives −1. None of the numbers then available (there wasn't yet a reason to call them "real numbers") would do this. So someone simply *imagined* one which would – a "mathematical fiction", as it was seen as at the time – and began to use it in their calculations. This was the complex number *i*. It wasn't until years later that someone realised *i* was part of a consistent number system which extended the familiar one and could be perfectly represented by the points in a plane.

Perhaps the simplest way of thinking about *i* is to imagine yourself walking along the number line towards zero, one step per counting number: 9, 8, 7, 6, 5, 4, 3, 2, 1, and when you get to 0... *you take a step to the right*. Stepping off the number line takes you into the weird world of complex numbers, which open up a whole new field of mathematical exploration. There's still a lot which isn't known about them – much mystery lurks in the complex plane. Here's an example of what some current "complex analysis" literature looks like[10]:

thus we find u on $X \smallsetminus D$ satisfying $\bar{\partial} u = \bar{\partial}(\chi f)$ such that

$$\int_{X \smallsetminus D} \frac{|u|^2 e^{-2\varphi}}{|z-x|^{2(n+s)}} \Omega_P(z) < +\infty.$$

Furthermore, $F = \chi f - u$ is holomorphic on $X \smallsetminus D$. It extends to the whole Ω: in fact, we only need to see that F is holomorphic in each variable $z_1, \ldots, z_p$ near D. We do it for z_1, using the classical approach: setting $z' = (z_2, \ldots, z_n)$, we write the Laurent series expansion $F(z_1, z') = \sum_{-\infty}^{+\infty} a_n(z') z_1^n$. As the integral $\int_{X \smallsetminus D} |F|^2 e^{-2\varphi} \Omega_P$ converges, the integral $\int_{X \smallsetminus D} |F|^2 \Omega_P$ converges too, and using Fubini's theorem, the integral $\int_{B_{\mathbb{C}}(0,\epsilon) \smallsetminus \{0\}} \frac{|F|^2}{|z_1|^2 \log^2 |z_1|} dV(z_1)$ is convergent. But then, thanks to Parseval's theorem:

$$\int_{B_{\mathbb{C}}(0,\epsilon) \smallsetminus \{0\}} \frac{|F|^2}{|z_1|^2 \log^2 |z_1|} dV(z_1) = C \sum_{-\infty}^{+\infty} |a_n(z')|^2 \int_{B(0,1)} \frac{|z_1|^{2(n-1)}}{\log^2 |z_1|} dV(z_1)$$

and necessarily, we have: $\forall n \leq -1, a_n(z') = 0$, which shows that $F(\cdot, z')$ admits an holomorphic continuation at 0.
Therefore, $F \in \mathcal{H}(X, \varphi)$, and as φ is has an upper bound near x, $f_x - F_x = u_x \in \mathscr{A}_x \cap \mathfrak{m}_{X,x}^{s+1}$, which concludes. □

Unfortunately, our sheaf $\mathcal{A}dj_D^0(\varphi)$ fails to coincide in general with the algebraic adjoint, as the next example shows:

Counterexample. *Let $X = (\mathbb{C}^2, 0)$, $\mathfrak{a} = \mathfrak{m}^6$, $H = \{z_1 = 0\}$, and $f(z_1, z_2) = z_1^2 z_2^3$. If $\varphi_\mathfrak{a} = 3 \log(|z_1|^2 + |z_2|^2)$ is a psh function attached to $\mathfrak{a}$, then we have:*

$$f \in \mathcal{A}dj_H(\varphi_\mathfrak{a}) \smallsetminus \text{Adj}(\mathfrak{a}, H).$$

The system of complex numbers has been known about for a couple of centuries, but the "reality" of it is only now starting to sink in (and it's still almost entirely unknown to the non-mathematical public).

One of the reasons why imaginary numbers are now considered to be no less "real" than the real numbers is their usefulness in science. There are strong connections with physical reality. Aerodynamic engineers designing wing shapes, *etc.* regularly use the mathematics of the complex plane. Even domestic electricians sometimes use *i* in their calculations (although they call it "*j*" to avoid confusion with the "*I*" which is used to denote current). The theory of AC electricity – that's the electricity produced by power stations and generators, as opposed to the DC electricity which is produced by batteries – depends on the use of *i*. So we wouldn't have AC electricity if we didn't have some grasp of complex numbers. To begin to understand why this is, we'd have to delve into the theory of *quantum physics*, the study of matter at the subatomic level (all electrical and electronic phenomena can ultimately be traced back to the "quantum" scale of reality). That theory involves some very sophisticated mathematics to which the complex numbers are absolutely essential.

Once you've learned a bit about this, you naturally come to accept that $\mathbb{C}$ is as "real" as $\mathbb{R}$ (whatever that means). If numbers are in any sense "built into the structure of reality" – and this is an ongoing matter of philosophical debate – then the complex numbers are as "built in" to it as the real numbers are.

Finally, you may have seen pictures like these:

They're called *fractals* and have been studied extensively by mathematicians since the 1980s[11]. Although elements of patterns like these show up all over the natural world, it has only been with the advent of sufficiently powerful computer technology that Western humanity has really been able to get a grip on this kind of geometry. Fractals such as those pictured above naturally inhabit the complex plane. Despite their infinitely detailed, intricate nature, to describe them precisely requires surprisingly little mathematics. There are simple formulas like "$z \to z^2 + c$" which, once you know how to interpret them, produce images like the ones above (and these images are infinitely detailed – you can zoom in indefinitely and they would keep revealing new structure, new details). But such formulas only make sense in the context of the complex plane – without access to complex numbers, we'd not be able to produce these images. It would be impossible to argue that these incredibly rich, detailed images are simply the products of human imagination (no one designed them, and no one can live to see more than infinitely tiny pieces of them, as they "go on forever"). So, the fact that it provides a natural "home" to such structures is yet another reason to see $\mathbb{C}$ as more than just some human creation or abstraction[12].

Chapter 17
The zeta function

Back to the story. We've heard a bit about a strange and wonderful creature called the "Riemann zeta function". The reason for that lengthy diversion through the number systems $\mathbb{N}$, $\mathbb{Z}$, $\mathbb{Q}$, $\mathbb{R}$ and $\mathbb{C}$ is that the zeta function's natural home is $\mathbb{C}$. Without $\mathbb{C}$, there's not much you can say about it, beyond the vague assertions on pages 2–3.

Unlike the various functions we encountered in Volume 1 which were "functions on $\mathbb{R}$", this is a "function on $\mathbb{C}$". Not only simple operations like adding, multiplying, *etc.*, but just about *any* kind of mathematical idea which applies to the real numbers can be somehow extended or generalised to the complex numbers. The idea of a function is a good example of this.

We can think of functions on $\mathbb{R}$ as taking points on the number line and transporting them to other points on the number line. We saw some examples back in Chapter 11. A function might, for example, take each number and multiply it by itself, or add 3 to it, or halve it, or divide it by its own logarithm, or send it to 7.261, or even just leave it where it is. A *graph* is the usual visual guide to what a function does to each real number. But I also introduced a more picturesque image which involved a "golfing sprite" which hits a ball located at any given number to the appropriate destination, in accordance with the function it represents. Admittedly it's a bit daft, but this visualisation, as opposed to the use of graphs, helps to dispel a widespread point of confusion – that is, functions aren't "pictures" as people sometimes mistakenly think. For our purposes, they're more helpfully thought of as something like *actions*.

The image of golfing sprites hitting balls from one point to another on $\mathbb{R}$ can be easily extended (or generalised) to $\mathbb{C}$, as $\mathbb{R}$ can be seen as just one line in the complex plane.

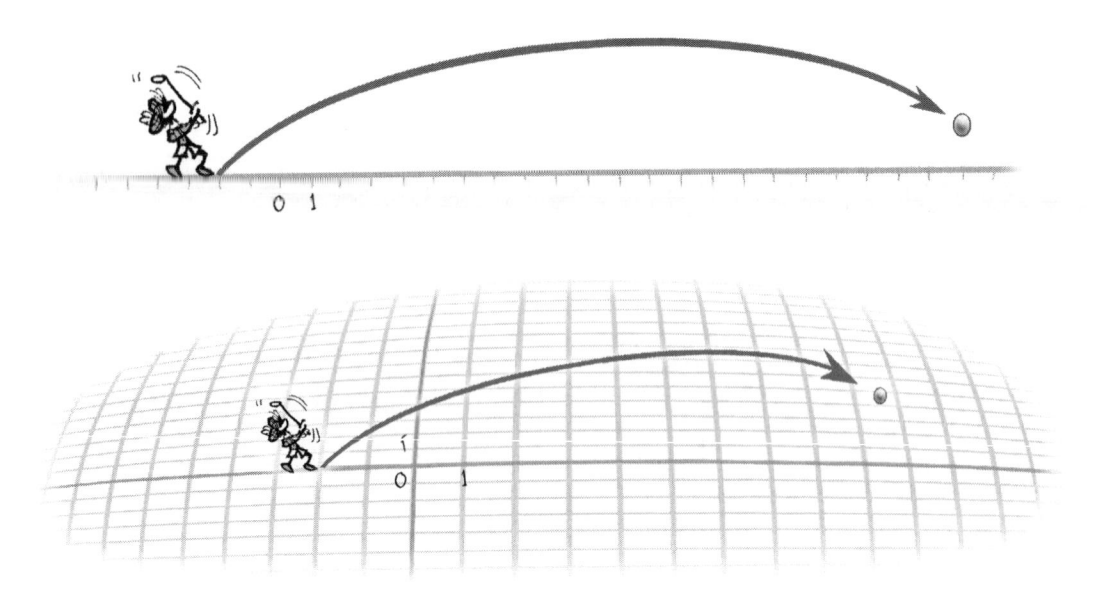

So, in this way, the idea of a function on $\mathbb{R}$ can be seen to extend (or generalise) to the idea of a function on $\mathbb{C}$.

A function on $\mathbb{C}$ (known as a "function of a complex variable" or just a "complex function") takes any given complex number, represented as a point on the complex plane, and transports it to some predetermined location in the complex plane – that is, to another complex number. Although I've said "another", in some cases, the destination can be the same as the location of the original point – that is, a function can leave some points where they are. These are called the *fixed points* of the function.

A given function is an unchanging, totally reliable mechanism, in that it will always "do the same thing" to a particular complex number. If we return to our golfing sprite imagery, each function is associated with a unique sprite which is entirely consistent in its behaviour – that is, if you place the ball in the same place several times, a particular sprite, when summoned, will hit it to exactly the same destination each time.

A very simple complex function which multiplies each number by itself does this:

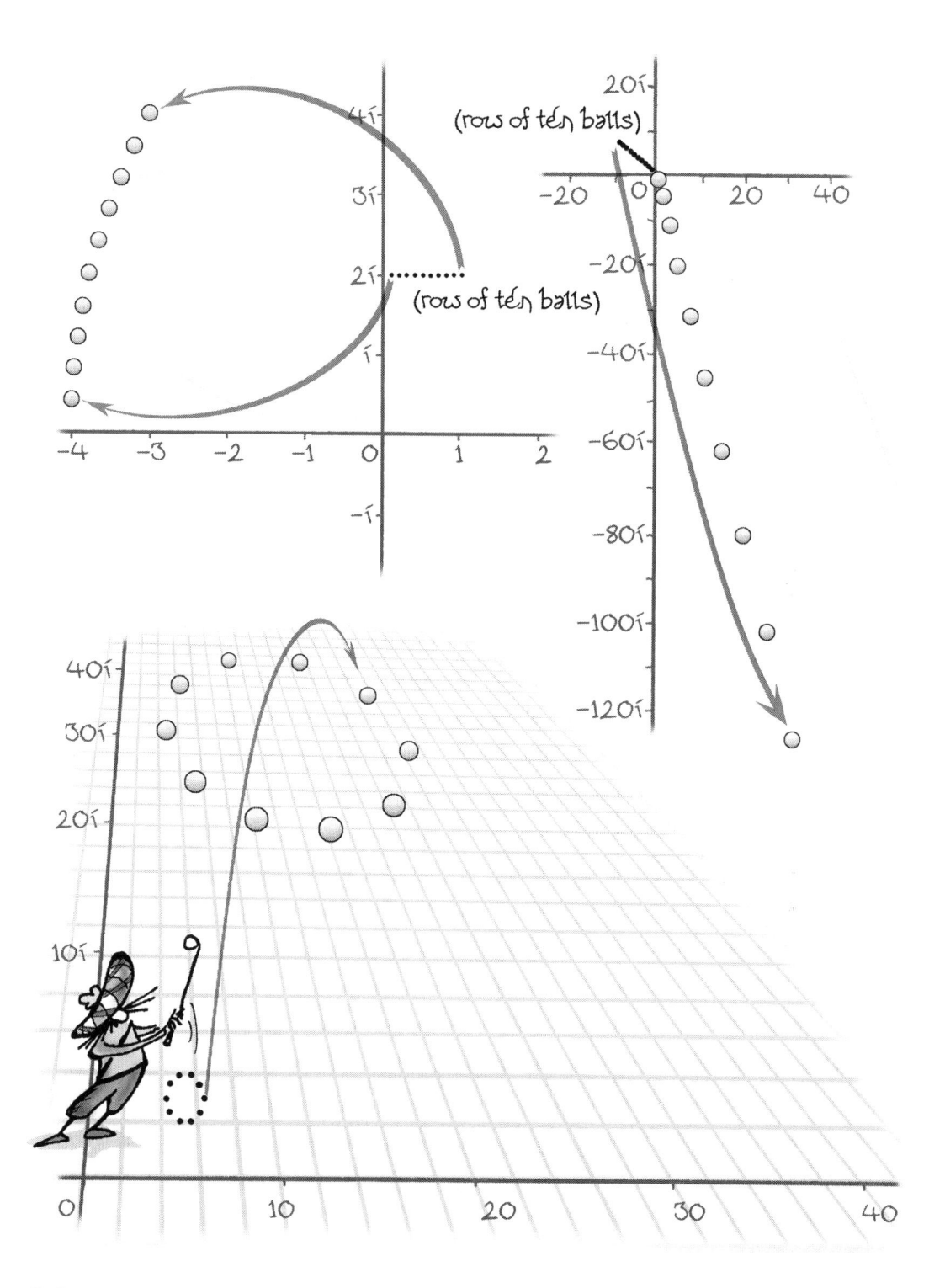

Only one or two examples are fully illustrated in each case. You should be able to deduce the rest.

31

One which multiplies by 2+3*i* and then adds 4*i* to the result looks like this:

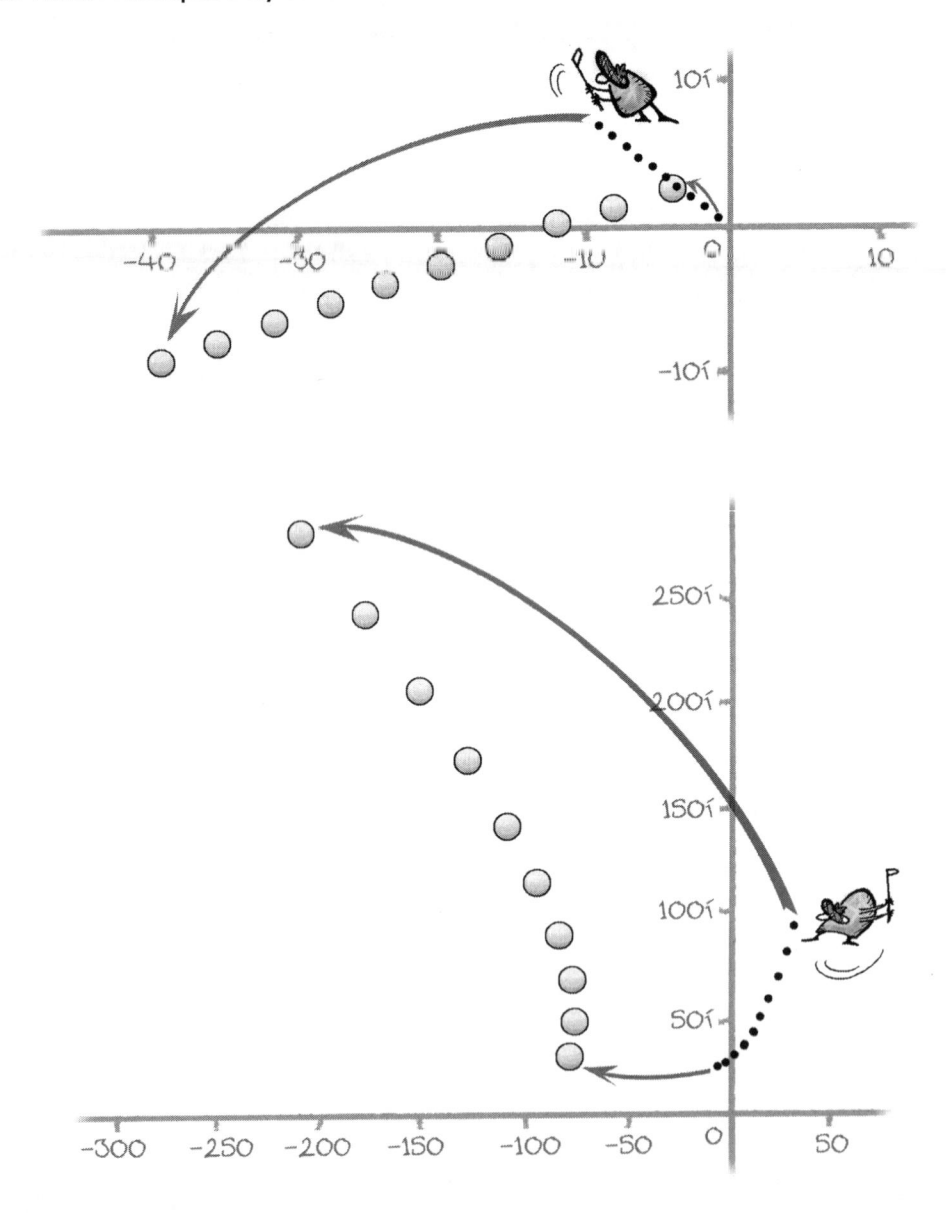

Sometimes a complex function is easy to describe geometrically, in terms of a "shift", a rotation, a magnification, or some combination of these. Sometimes this is just not possible – but it is possible to work with functions mathematically without necessarily being able to describe a clear rule for their behaviour in anything like informal language. There's a whole field of complex function theory with its own specialised language and notation, as we saw in the (possibly quite alarming) example back on page 26. Fortunately, though, there's only one complex function which is really going to concern us: the Riemann zeta function.

As we've seen, the "golfing sprite" visualisation carries over very easily from $\mathbb{R}$ to $\mathbb{C}$. There's not much difference between hitting balls from one part of a line to another or hitting them from one part of a plane to another. The idea of a *graph* representing a function does *not* carry over very well, though. The main reason for this is that the number line is 1-dimensional and we need two copies of it (one horizontal, one vertical) to make a graph, so the graph is 2 (or 1+1) dimensional. In a similar way, the complex plane is 2-dimensional, so if we were to make a graph of a complex function, it would have to be 4 (or 2+2) dimensional. Because we're physically stuck in three-dimensional space, it's not possible to draw 4-dimensional graphs. There are some clever tricks which allow you to graph *some aspects* of a complex function – these often require the use of 3-dimensional perspective drawing or colour. But they can never depict the full behaviour of the function (as ordinary $\mathbb{R}$ function graphs can so easily do), so these are not really satisfactory. This difficulty explains why the golfing sprites were introduced in the first place. Often, graphs are emphasised to the extent that many students confuse them for the functions they represent. Our visualisation takes this emphasis away, and it clearly shows how the idea of functions can be naturally extended from $\mathbb{R}$ to $\mathbb{C}$.

As before, you insert your tee at a chosen point on the complex plane golf course, place the ball and summon the appropriate function sprite. It then appears and hits the ball to another point on the plane. Remember, the important thing is that if you choose the same point again, a particular sprite will *always* hit it to the same resting place. In this sense, these sprites are completely predictable and reliable, their behaviour embodying the "behaviour" of a particular function. By placing enough balls in well-chosen locations and paying attention to where a given sprite hits them, it would be possible to build up some knowledge of the function and its various properties.

Some functions are *continuous*. As with functions in $\mathbb{R}$, this means roughly that if you place two balls close together and then summon the appropriate function sprite to hit them both, they'll be sent to nearby locations (see Chapter 11). Although that's just a vague description, it will do for our purposes – mathematicians use an ultra-precise definition. Not all functions are continuous, in fact most aren't, but the more widely studied ones tend to be.

Mathematicians study functions (both real and complex) because they can often be in some way applied within the physical world, or used as tools for further mathematical exploration. But they're also studied as interesting objects in their own right. There are "special functions", with all kinds of properties that get described as "curious", "beautiful", *etc.*, and "families" into which functions can be classified (with names like "elliptic functions", "theta functions" and "*L*-functions") much as plant or animal species might be classified into families.

If you're finding any of this a bit confusing, don't worry. There's only one very special function on the complex plane which we're going to concern ourselves with, so just hang on to the image of a lone golfing sprite "doing its thing" on the endless, flat golf course.

I've mentioned that functions (both on $\mathbb{R}$ and on $\mathbb{C}$) can have significant points associated with them called "fixed points". Those are points which the function leaves where they are (it transports the point in question to...itself). But there's another type of significant point which is often studied. These are called the *zeros* of the function and they're simply points which the function transports to 0. When working with a function on ℝ, we can draw its graph, and the zeros of the function are easily identified from this graph – they're the points where it crosses the number line.

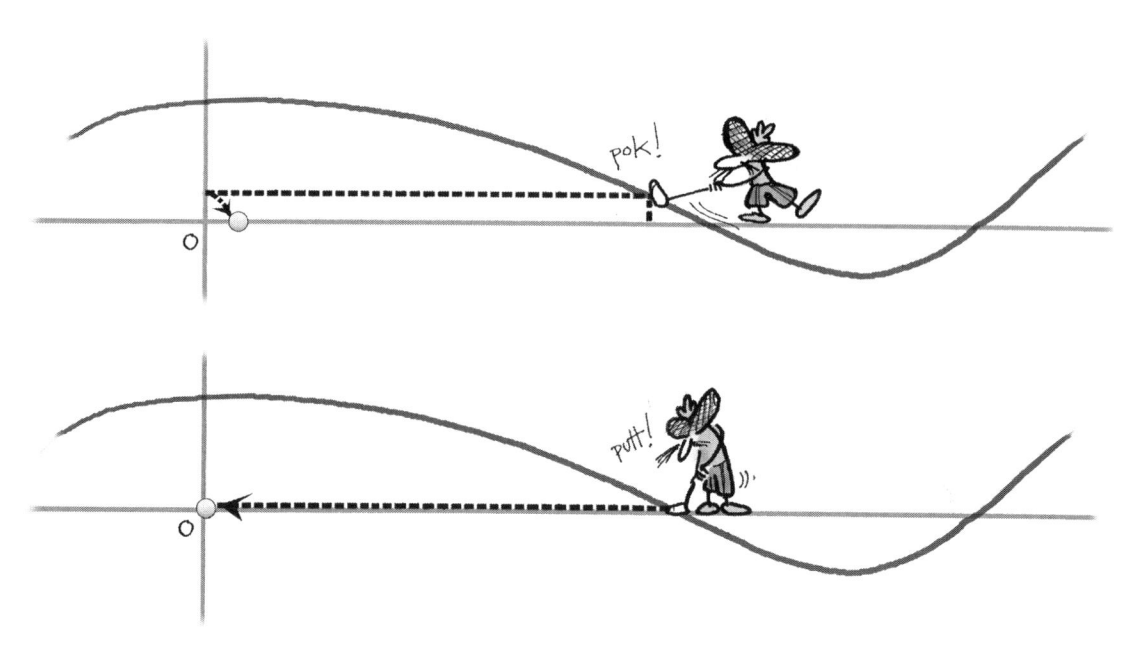

Recall the visualisation we used in Chapter 11: the ball is hit vertically, turns a corner when it hits the graph, then does a quarter clockwise rotation when it hits the vertical axis. If the graph passes through the horizontal axis at the relevant point, no vertical displacement or rotation is necessary, so the ball stops at the vertical axis, that is, at 0.

As I've said, it's not really possible to draw graphs of functions on ℂ, so the zeros of such functions can't be identified in this way. However, using the golfing sprite imagery, we can develop a very clear picture of what a zero of a complex function would be. If the complex plane is to be treated as an infinite, flat golf course, then we introduce a single hole at 0, where the real and imaginary axes cross. If you place the ball at a point and a particular function-sprite hits it into the hole at 0, then that point is a zero of the function in question.

The zeros of a function (on ℝ or on ℂ) are generally of great importance when studying its "behaviour" in a broad sense. Given a function, one of the first things a mathematician would think to do would be to determine where its zeros are located, since so much else can be deduced from this information.

It's also possible to discuss functions which are defined not on the *whole* of $\mathbb{R}$ or $\mathbb{C}$, but just on *some part* of $\mathbb{R}$ or $\mathbb{C}$. So, for example, you can have a function which is defined only on positive real numbers (that part of $\mathbb{R}$ "to the right of zero"), or on all of $\mathbb{R}$ except at 0, or for all real numbers greater than 1. Similarly, you can have (for example) functions defined only on that part of $\mathbb{C}$ above the real axis, or only on that part of $\mathbb{C}$ to the right of the imaginary axis, or on all of $\mathbb{C}$ except the point $1 + 3i$. In these situations, we'll just have to decree that placing the ball anywhere else is not allowed. Alternatively, we could imagine the function sprite appearing and hitting balls placed at such "disallowed" locations off into space so that they have no destination on the number line plane.

In the 1740s, Leonhard Euler was studying a particular function which is defined on part of $\mathbb{R}$, particularly on those real numbers greater than 1. Before describing how his function works, it would be helpful to explain why he was interested in it.

If you could add an infinite number of things together, you might expect the result to be infinite in size, but this isn't always the case. Certainly, if we add

$$1 + 1 + 1 + 1 + 1 + 1 + 1 + 1 + 1 + 1 + 1 + 1 + \cdots,$$

then we're not going to end up with something finite (any finite number will be eventually surpassed by adding enough 1's). But what about

$$\tfrac{1}{2} + \tfrac{1}{4} + \tfrac{1}{8} + \tfrac{1}{16} + \tfrac{1}{32} + \cdots \,?$$

Notice how each fraction being added is half the size of the previous one.

There is a very well-established mathematical framework in which this "infinite sum"

$1/2 + 1/4 + 1/8 + 1/16 + 1/32 + \cdots$ can be understood to produce the finite result 1:

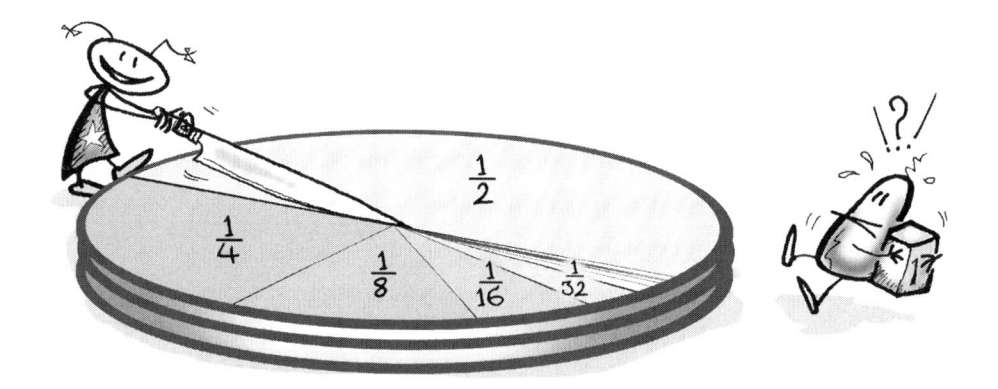

$$1/2 + 1/4 + 1/8 + 1/16 + 1/32 + \cdots = 1.$$

So some "infinite sums" produce finite results and others don't.

Euler had been looking at the sum $1 + 1/2 + 1/3 + 1/4 + 1/5 + 1/6 + 1/7 + \cdots$. He was able to prove that this does *not* produce a finite result. Even though the amounts you're adding are getting smaller and smaller, it's possible to prove that the sum will *eventually* surpass any finite number you choose. So the result is infinite.

However, Euler noticed that if you multiply each of the numbers by itself...

$$1 \times 1 = 1, \quad 1/2 \times 1/2 = 1/4, \quad 1/3 \times 1/3 = 1/9, \quad 1/4 \times 1/4 = 1/16, \quad 1/5 \times 1/5 = 1/25, \cdots$$

...and add the results together...

$$1 + 1/4 + 1/9 + 1/16 + 1/25 + \cdots$$

...then that *does* give a finite result. Similarly, if you multiply them by themselves again...

$$1 \times 1 \times 1 = 1, \quad 1/2 \times 1/2 \times 1/2 = 1/8, \quad 1/3 \times 1/3 \times 1/3 = 1/27,$$
$$1/4 \times 1/4 \times 1/4 = 1/64, \quad 1/5 \times 1/5 \times 1/5 = 1/125, \cdots$$

...and add them together...

$$1 + 1/8 + 1/27 + 1/64 + 1/125 + \cdots,$$

then that also gives a finite result.

The numbers just keep getting smaller, so this works for any "power" (any number of times you multiply each of the fractions by itself).

If, for each counting number after 1 we imagine the logarithmic spiral (remember, these are also called *equiangular* spirals) of that "base" (see Chapter 9)...

Recall that if a logarithmic spiral centred at O on the number line passes through 1, then its next crossing gives its base. Here we see spirals of base 2, 3, 4, 5 and 6.

...then 1/2, 1/3, 1/4, 1/5, ... turn out to be the points on the number line where these spirals make their first "inward" crossings from 1:

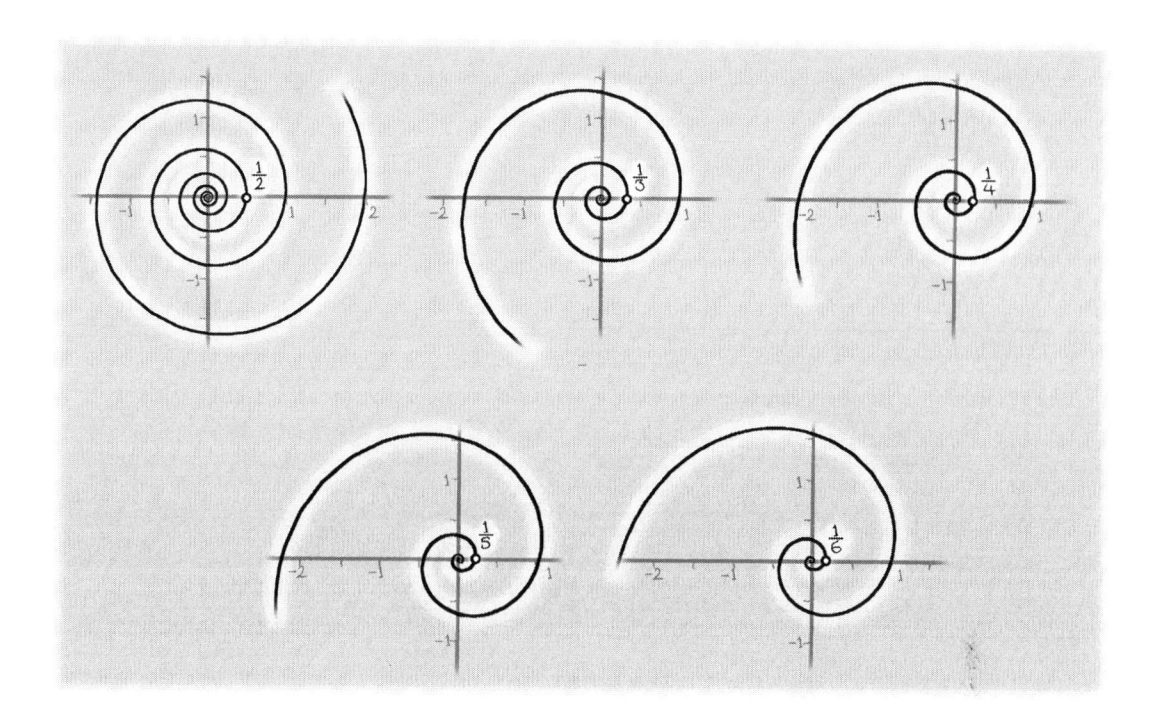

Similarly, $1/2 \times 1/2 = 1/4$, $1/3 \times 1/3 = 1/9$, $1/4 \times 1/4 = 1/16$, $1/5 \times 1/5 = 1/25, \ldots$ turn out to be the points where they make their second "inward" crossings:

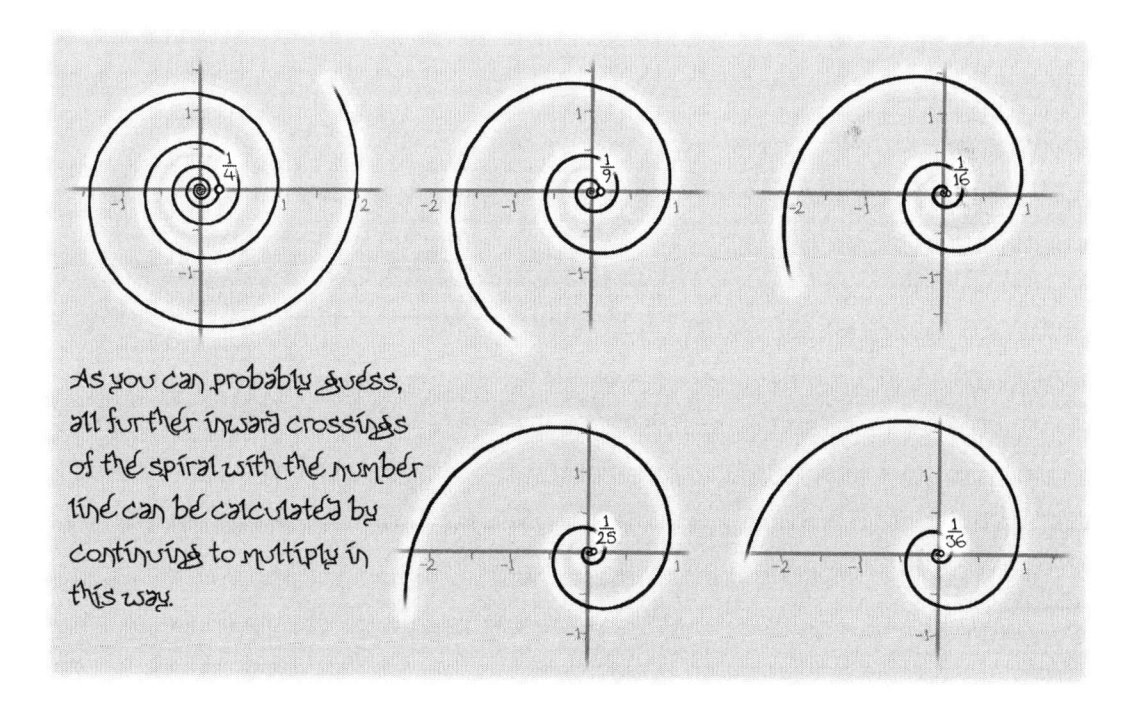

As you can probably guess, all further inward crossings of the spiral with the number line can be calculated by continuing to multiply in this way.

In fact, if we pick *any* number of "inward coils" (not just a counting number) bigger than 1, say 1.17, and mark the relevant points on our sequence of spirals,...

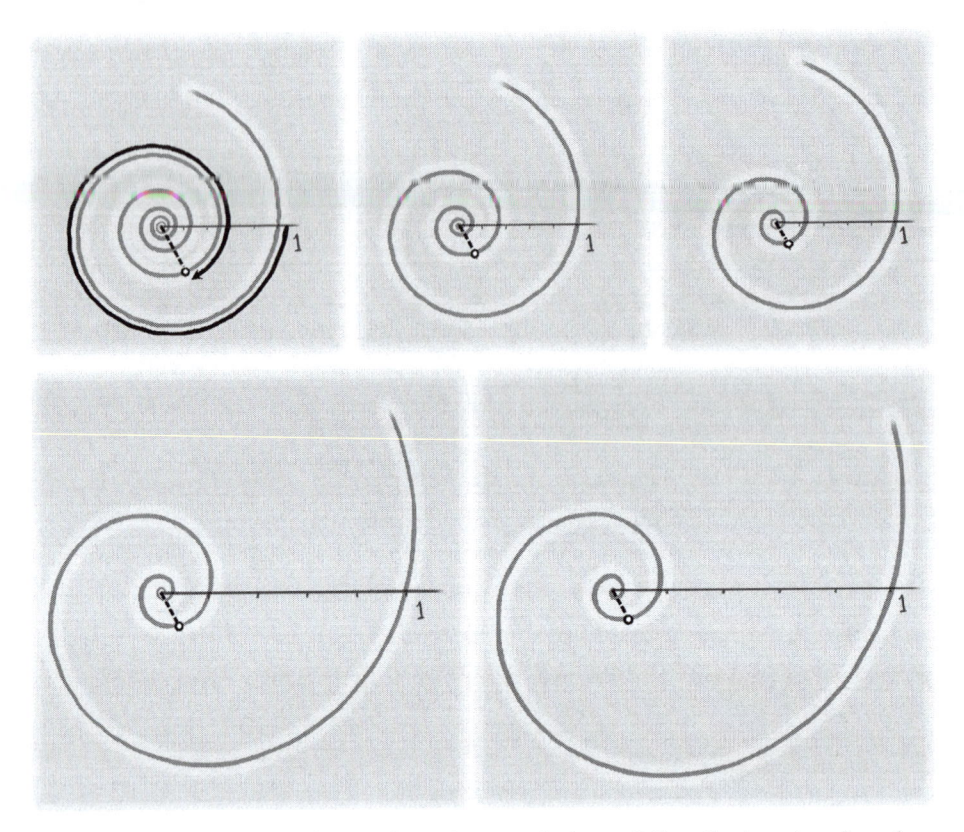

Imagine yourself travelling clockwise from 1, completing a full coil, then just less than a fifth of a second coil. Mark the point where you stop.

...measure the distance to zero from each of these (in this case, we get approximately 0.44442, 0.27654, 0.19751, 0.15212, 0.12290,...) and add all of these to 1[1] (giving approximately 1 + 0.44442 + 0.27654 + 0.19751 + 0.15212 + 0.12290 + ...), then *we always get a finite sum*. In this case, it's 6.471805... The greater our chosen number of "inward coils", the closer all of the resulting points will be to zero, so the closer the resulting sum will be to 1. The closer our chosen number of inward coils is to 1, the closer the resulting points will be to 1/2, 1/3, 1/4, 1/5,..., so the closer the finite sum will be to an infinite sum (as we would get if we actually chose 1) – *but it will still be finite*. Here's a table to give you some sense of how this works:

40

number of inward coils	sum of 1 plus all of the distances-from-centre
100	1.00000000000000000000000000000078886
50	1.0000000000000000888178
10	1.000994577
5	1.036927755
2	1.644934068
1.5	2.612375349
1.3	3.931949212
1.1	10.584444846
1.05	20.58084430
1.001	100.57728847

This allows us to define a function on part of ℝ, particularly that part of ℝ to the right of 1. Given any such real number, we can think of that many inward coils, which defines a point on each of our sequence of spirals, each of which is at a finite distance from zero. Adding all of these distances together (plus 1), we get a finite number, and this is where the function transports our originally chosen number. So 5 gets transported to 1.036927755..., 1.3 gets transported to 3.931949212..., 1.001 gets transported to 100.5779433..., and so on, but the function *is not defined at 1*. The (seemingly unremarkable) graph of this function looks like this:

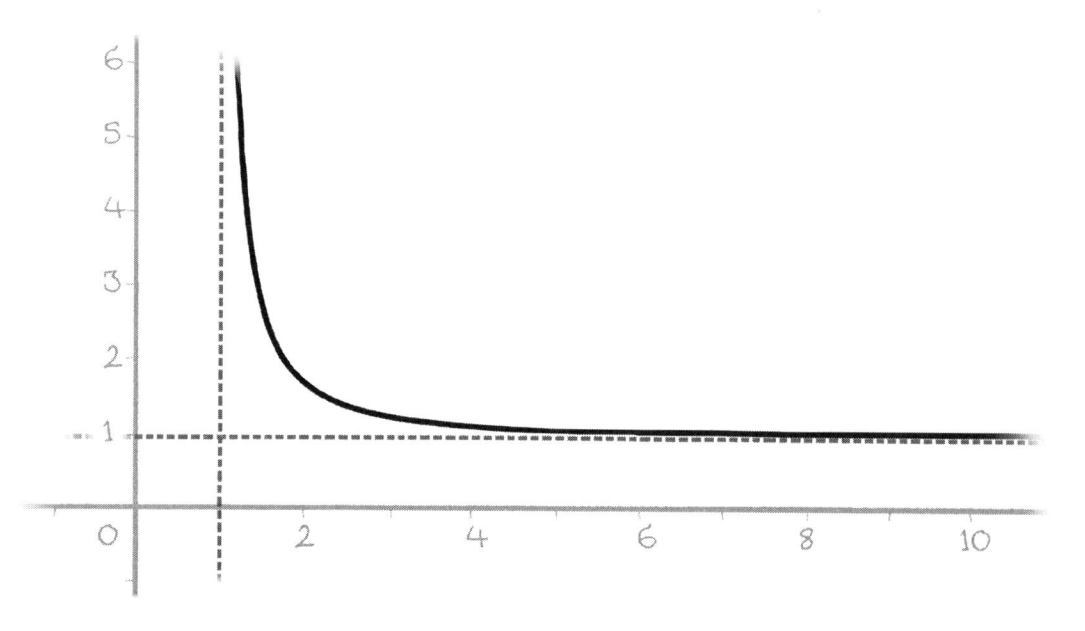

Remember that for each point on the number line (to the right of 1), the height of the graph above the point tells us where that point is to get transported. So the closer a point is to 1, the farther away it's going to get "hit", and the farther out along the number line it is, the closer it'll get hit to 1. There's a unique fixed point at 1.8337726516... (the function leaves it where it is). This is a kind of "pivot": every other point where the function is defined gets transported from one side of it to the other. So, in a very loose sense, this function can be thought of as turning part of the number line "inside out".

Euler originally denoted this function by the Greek letter ζ (written "zeta" in English), so it's sometimes called the *Euler zeta function*.

chapter 18
back to prime numbers

It's been a while since there's been any mention of prime numbers, but fear not, they're about to come back into the picture.

Euler showed that the zeta function he was studying can also be defined using the sequence of primes rather than the sequence of counting numbers (as we've just done). It works like this. Rather than imagining a spiral for each counting number, we imagine a spiral for each *prime* number:

Here we see logarithmic spirals with bases 2, 3, 5, 7 and 11. All crossings beyond 1 will be powers of the prime number in question.

Then, for any chosen real number bigger than 1 (let's continue to use 1.17), we again find the point that many "coils inward" from 1 (towards 0) on each spiral.

43

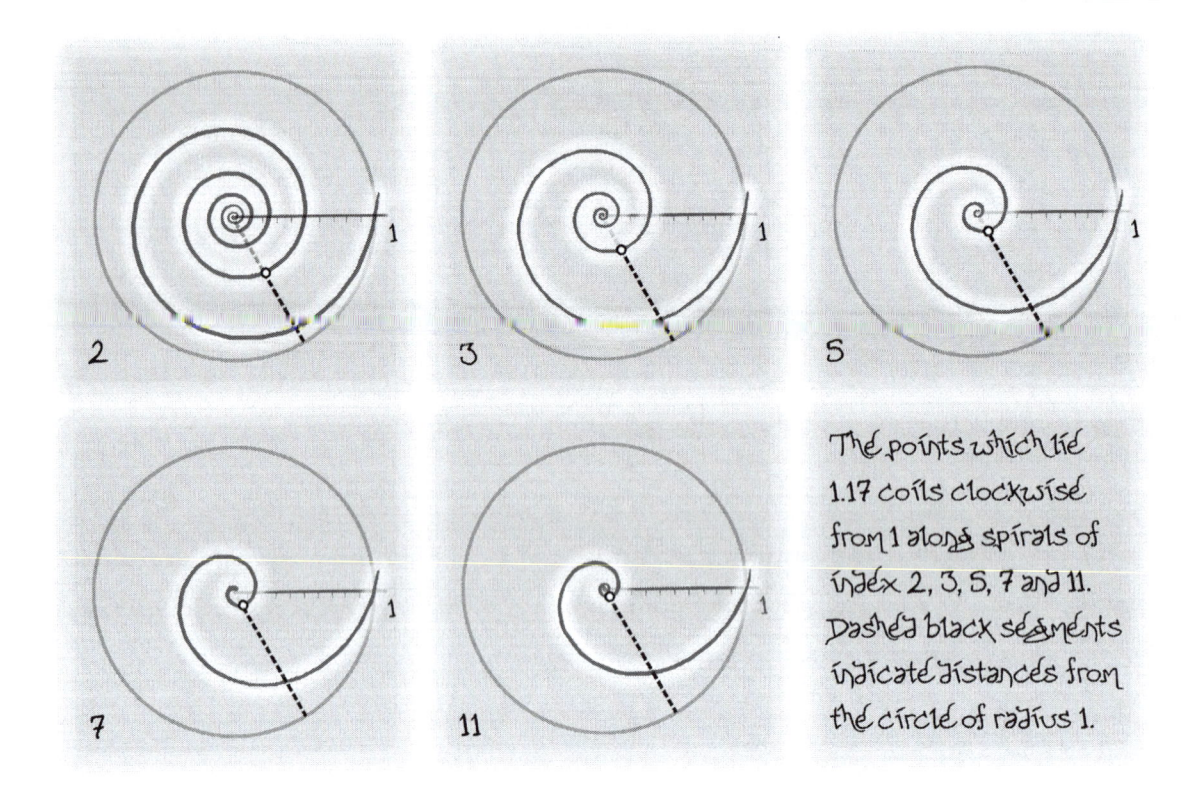

The points which lie 1.17 coils clockwise from 1 along spirals of index 2, 3, 5, 7 and 11. Dashed black segments indicate distances from the circle of radius 1.

Each of these points is at some finite distance (shown as a dashed black line segment) from the circle of radius 1 which is centred at the centre of the spiral. Take all of those distances: 0.55557..., 0.72345..., 0.84787..., 0.89737..., 0.93952..., 0.95026..., ... There are infinitely many of these numbers (one for each prime), but this time, rather than adding them all together, we're going to *multiply* them all together. Just as you can sometimes add infinitely many numbers together and get a finite result, you can also sometimes *multiply* infinitely many numbers together and get a finite result. In this case, the result we get is 0.154516... We then *divide 1 by that*. This will give us the same result as with the first approach.

In order to make sense of dividing 1 by a number like 0.154516... there are a couple of useful visualisations, the first involving our elastic rope. To find 1 divided by 0.154516..., we mark our rope at 0.154516... and 1, and then pull it until the "0.154516..." mark reaches 1. Where the "1" mark ends up is 1 divided by 0.154516...

Alternatively, we can think of dividing 1 by a number (say 0.27708) in this way:

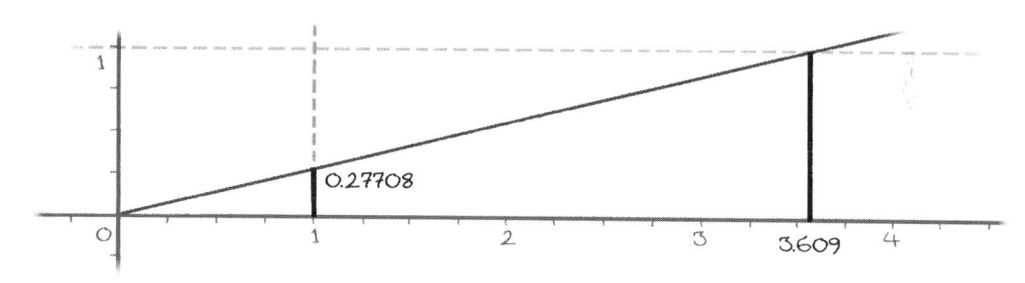

Create an upright at 1, whose height is the number in question, then take a line from 0 through the top of this and note where it crosses the horizontal line of height 1.

The details of what we've recently seen won't concern us too much. The key things to retain here are:

✯ Euler was looking at a function which is defined on part of $\mathbb{R}$ (specifically, for all real numbers greater than 1).

✯ This function can be described in terms of spirals in two separate ways:

   ✯ One of these involves the sequence 1, 2, 3, 4, 5, 6,... and the *addition* of infinitely many quantities.

   ✯ The other involves the sequence 2, 3, 5, 7, 11,... and the *multiplication* of infinitely many quantities.

45

The golfing sprite associated with the Euler zeta function will hit balls from near 1 to distant regions of $\mathbb{R}$ and *vice versa*, but it does so in a very specific way. We could imagine it using either of the two procedures recently described in order to determine where to hit the ball. The important thing here is that these two different procedures – one involving all of the counting numbers and addition, the other involving just the primes and multiplication – *give the same result*. This fact, although well understood, is often described as "remarkable" and is expressed by mathematicians as the *Euler product formula*:

$$\sum_{n=1}^{\infty} \frac{1}{n^x} = \prod_{p} \frac{1}{1-p^{-x}}$$

Euler proved this formula. A description of how is given in Appendix 10 if you're interested (as with all of the appendices in this volume, you'll need some familiarity with mathematical notation and reasoning). The proof is very heavily dependent on the Fundamental Theorem of Arithmetic (FTA), that result involving the uniqueness of prime factorisations which we met back in Chapter 4. In fact, the Euler product formula is sometimes described as a kind of extension or "analytic version" of the FTA – the same basic truth, but in a wider, "analytic", context. "Analytic" here refers to *analysis*, that branch of mathematics which deals with such things as those infinite sums we've been looking at, and the more general idea of a *limit* (calculus is the part of analysis which the non-mathematical public tends to have heard of). Euler's great achievement in proving his product formula was to build a bridge between number theory and analysis, two previously separate fields. In this way, he helped give birth to a new branch of mathematics which is now known as *analytic number theory*.

Just to give you a feel for how the FTA comes into this, recall that it says that if you multiply primes together in all finite combinations, then you'll get each counting number exactly once. If you understood the notation and started experimenting with the formula illustrated opposite, you'd find that the right-hand side "expands out" into something involving all possible finite combinations of primes, each combination being multiplied together. For it to equal the left-hand side, it's necessary that each counting number can be arrived at in this way, and arrived at by means of *one and only one* combination of primes. For this reason, we can say that Euler's product formula is true precisely because the Fundamental Theorem of Arithmetic is true.

So Euler's zeta function (which the product formula shows in two equivalent forms) relates the whole set of counting numbers to the whole set of primes in a way which many people find initially surprising or remarkable, even though it's not actually *that* difficult to see why it's true.

Studied in depth, the formula reveals itself as a sort of embodiment of the relationship between the system of counting numbers and the sequence of prime numbers, and of the relationship between addition and multiplication. It allowed new methods of mathematical investigation to be used in exploring these relationships. And it allows us to learn things about the mysterious sequence of primes by studying the (much simpler) system of counting numbers.

We must make a clear distinction between the Euler *zeta function* and the Euler *product formula*. The product formula is a mathematical assertion, which was proved to be true. It states that two different things (versions of the zeta function) are equal. The zeta function is *not* a mathematical assertion, it cannot be true or false, it just *is* – it's a *function*, something which we can think of as transporting points to other points on part of the number line $\mathbb{R}$ in a very specific way.

Looking at what the Euler zeta function "does" on $\mathbb{R}$, its behaviour doesn't seem particularly exciting. It has one fixed point, but no zeros. It's graph is a simple looking curve which doesn't suggest that this function has any surprises in store for us. But the properties which we've seen certainly allow it to qualify as a "special function" and, because of its properties, it was studied into the 19th century (up to Riemann's time).

One curious fact about Euler's zeta function is where it transports the number 2. If you go back to the graph, you'll see that it's to a number a bit bigger than 1.6. It turns out to be 1.64493..., which is what you get if you multiply π (3.14159...) by itself and then divide the result by 6. So π is involved somehow[1].

If you look at where the Euler zeta function transports other counting numbers, you'll often find other quantities which involve π in a similar way.

This number 1.64493..., which you get by multiplying π by itself and dividing by 6, is also relevant to prime factorisation. It has been proved that if you pick two counting numbers at random[2], the probability that their factorisations will have no prime factors in common is *1 in 1.64493...* Another way of expressing this is as a percentage, which is found by dividing 1 by 1.64493... (which gives about 0.608) and then multiplying that by 100, to give approximately 60.8%. That tells you how often two randomly chosen counting numbers have no common prime factors (they're called *relatively prime* in that case) – approximately 60.8% of the time. This gives us a small hint as to the role that the Euler zeta function plays in "encoding number theoretical information" and why it would continue to be studied for centuries after Euler discovered it.

By Riemann's time, mathematicians were becoming very familiar with the workings of the complex plane. So, sooner or later, someone was going to attempt to extend Euler's zeta function from its original home on part of the number line $\mathbb{R}$ to the complex plane $\mathbb{C}$. It was Riemann who succeeded in doing this.

What would it mean to "extend" the Euler zeta function? This would have to involve a function on $\mathbb{C}$ whose behaviour exactly matched that of the Euler zeta function on that part of the real axis to the right of 1 (or at least a function on *some part* of $\mathbb{C}$ which includes that part of $\mathbb{R}$ to the right of 1 and which coincides with the Euler zeta function there). In other words, if you placed a golf ball at some point on that part of the real axis and then summoned the appropriate (complex) function sprite, it would hit it to exactly the same destination as the Euler zeta function sprite would. Any such complex function could be said to extend Euler's function (agreeing with it wherever it was defined, but also being defined elsewhere).

Riemann not only found a way to extend the Euler zeta function to $\mathbb{C}$ (actually, to all of $\mathbb{C}$ except the point 1), but also found a *uniquely natural* extension. The resulting complex function is, as you've probably guessed, the "Riemann zeta function" which was informally described on pages 2–3. But what do I mean by "uniquely natural"?

Mathematicians studying complex functions are particularly interested in the family of these which share the property of being *analytic* (also sometimes called *holomorphic*). We won't go into the details of what "analytic" means precisely (it's not important for our purposes), but analytic functions have some very special properties, perhaps the most remarkable of which is this:

The behaviour of the whole function can be determined from its behaviour on any tiny piece of $\mathbb{C}$.

This means that if you knew what you were doing and had unlimited time, by experimentally placing golf balls in just one tiny piece of the $\mathbb{C}$ golf course (as tiny as you like) and studying where the appropriate analytic function sprite hit them, you could accurately predict where it would hit a ball from *any* location on the plane.

Most of the really interesting complex functions are analytic. Riemann's zeta function can be shown to be the *only* analytic function which extends Euler's zeta function to $\mathbb{C}$ (minus one point). And so, because of the central status of analytic functions in complex analysis, the extension could be described as "uniquely natural".

It's as if the behaviour of Riemann's zeta function was already there, "hidden inside" Euler's, just waiting for a chance to "express itself" by being placed in the context of $\mathbb{C}$...as if it were the genuine article which had been there all along while people were stuck with studying just one little slice of its behaviour (on part of $\mathbb{R}$) which seems fairly unremarkable. Euler was unable to extend his function because the mathematics of the complex plane wasn't available to him in the 1740s. It was inevitable that someone was eventually going to make this leap once complex numbers became part of mathematical discourse, although this shouldn't be seen as detracting from Riemann's genius.

I've explained why you can't draw a graph of a complex function, but there *are* various ways in which you can use their behaviours to generate images. The images below have all been generated in different ways using the Riemann zeta function. They should give you some sort of feel for the strangeness of this beast.

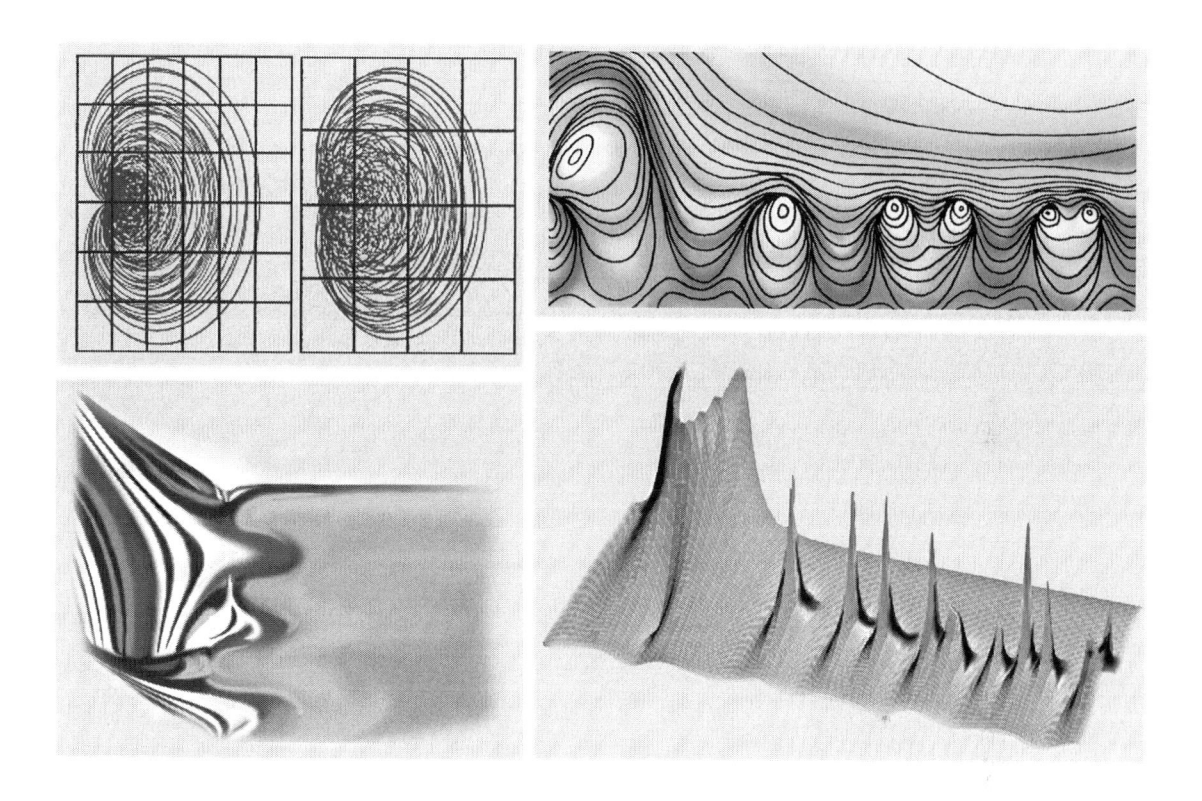

With these mysterious visual impressions in mind, let's have a closer look at how this thing actually "behaves".

If you choose complex numbers ever closer to 1 (approaching it from any direction you like in the plane), you'll find that our Riemann zeta function sprite will hit the placed balls farther and farther out (but always to specific locations, of course). You can get the sprite to hit the ball as far as you like if you place the ball sufficiently close to 1, but you're not allowed to place it *at* 1. That is, the Riemann zeta function is *not defined* at 1. If you really must insist on placing it there, the sprite would hit it off the course altogether, "out to infinity" – but infinity can be a tricky concept, so we'll stick to our simple prohibition that you can't place the ball there. Mathematicians describe the point 1 as a *pole* of the Riemann zeta function, a singular point in the sense that the North and South Poles of the Earth are. So, in the spirit of this terminology (and *Winnie-the-Pooh*[3]), we'll imagine a (more mundane) pole sticking out of the complex plane golf course at this point, which conveniently stops us from being able to place our ball there.

The Riemann zeta function is known to be a *continuous* function on the complex plane. You might recall that this means its sprite will always hit balls placed close together to nearby destinations. So, we can place lines and rings of balls in certain places on the plane and then, once the sprite has been summoned to hit them all, look at how these shapes get transformed:

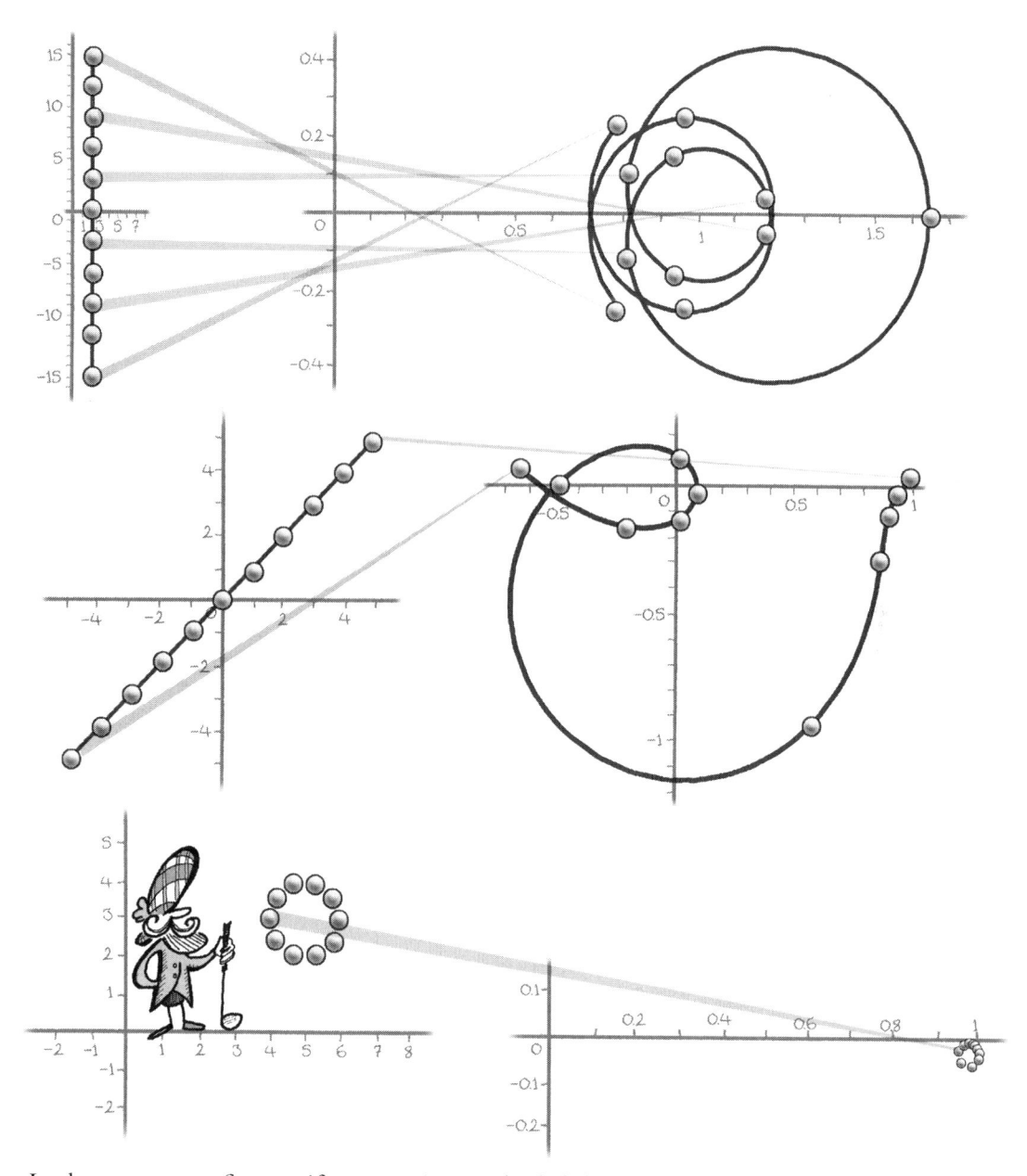

In the upper two figures, if *every* point on the left-hand line segment had a ball placed on it, the entire right-hand curve would be produced by the sprite's action.

And because Riemann's zeta function is an extension of Euler's, placing the ball anywhere on the real axis to the right of the pole at 1 will result in the Riemann zeta sprite hitting it to the same point as would his older brother, the Euler zeta sprite.

Another notable feature of the Riemann zeta function is that if you try to drop the ball directly into the hole (that is, you try to place it at 0), it refuses to go in. Instead, the function sprite immediately appears and taps it over to the point at −1/2:

This is simply because the zeta function transports the complex number 0 to the complex number −1/2 (recall that the Euler zeta function isn't even *defined* at 0, but the Riemann zeta function is its extension, defined *everywhere* on the complex plane except the point 1). This means that 0 is not a zero of the Riemann zeta function. If it were, the function sprite would leave it alone when it was placed there (functions transport their zeros to 0, remember).

We've seen that the Euler zeta function doesn't have any zeros so we can be sure that the Riemann zeta function also has no zeros on that part of the real axis to the right of 1. But does it have any zeros anywhere else? As I've said, finding a function's zeros is one of the first things a mathematician would think to do, and they turn out to be *hugely* important in the case of the Riemann zeta function as we shall go on to see. So we shall imagine a game involving the Riemann zeta sprite where the objective is to *get the ball into the hole*. That is, the idea is to find points on the complex plane which get transported to 0. When you place a ball at such a point and summon the Riemann zeta function sprite, it will get hit directly into the hole.

If we experimented with the placing of balls at various "obvious" locations, we'd eventually find that balls placed at −2, −4, −6, −8, ... (the negative even integers) all get hit into the hole. So −2, −4, −6, −8, ... are all zeros of the Riemann zeta function.

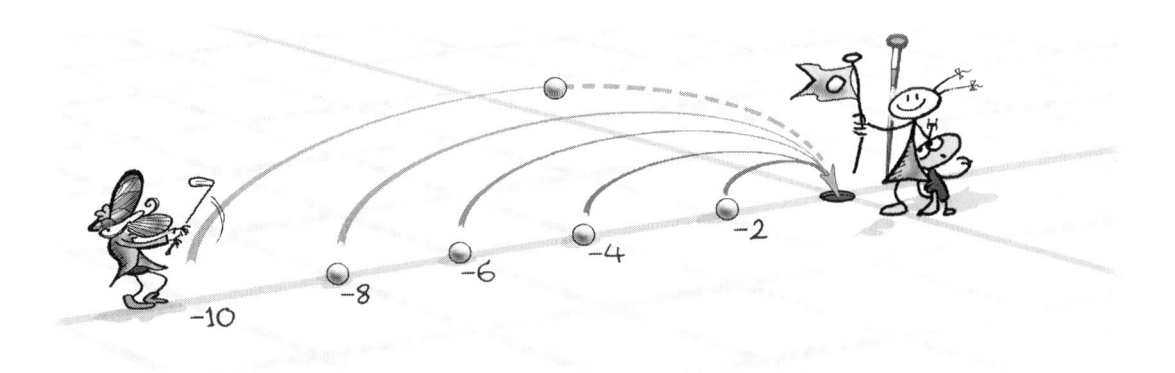

Given an infinite supply of time to place balls at the other "obvious" grid points along the real and imaginary axes, you'd eventually convince yourself that there are no other zeros to be found in these places.

I've claimed that the Riemann zeta function is the unique analytic function which extends the Euler zeta function. That's sufficient to identify it among complex functions, as far as mathematicians are concerned. But how does it actually *work*? When we place a ball at a general location on the plane (that is, somewhere other than on that part of the real axis where Euler's zeta function is also defined), how does the function sprite know where to hit it? For Euler's function, we saw a calculation involving spirals and the adding (or multiplying) of lots of little distances of points on them from 0 (or from a circle of radius 1).

There's a similar procedure we can use to calculate the behaviour of the Riemann zeta function — that is, to calculate the destinations of balls placed at given points — but it only works on half of the plane. If we draw a vertical line through the pole at 1, this procedure will only work at points to the right of that line. As I'll explain later, though, there are "tricks" by which we can then extend this to the whole plane.

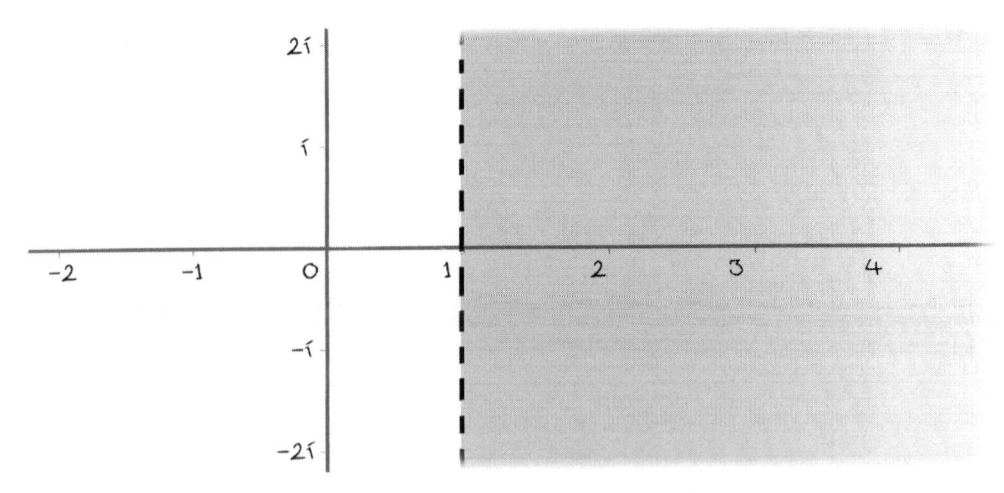

The "right half-plane" on which our initial definition of the Riemann zeta function will be valid.

We already know how to calculate what the Euler zeta function does on that part of the real axis in this half-plane, so we'll only describe the procedure for points above or below it. Any complex number in either of these regions can be used to define a logarithmic spiral, as follows. We simply draw a line from 0 through the point in question, then use the angle that this makes with the real axis to define our spiral:

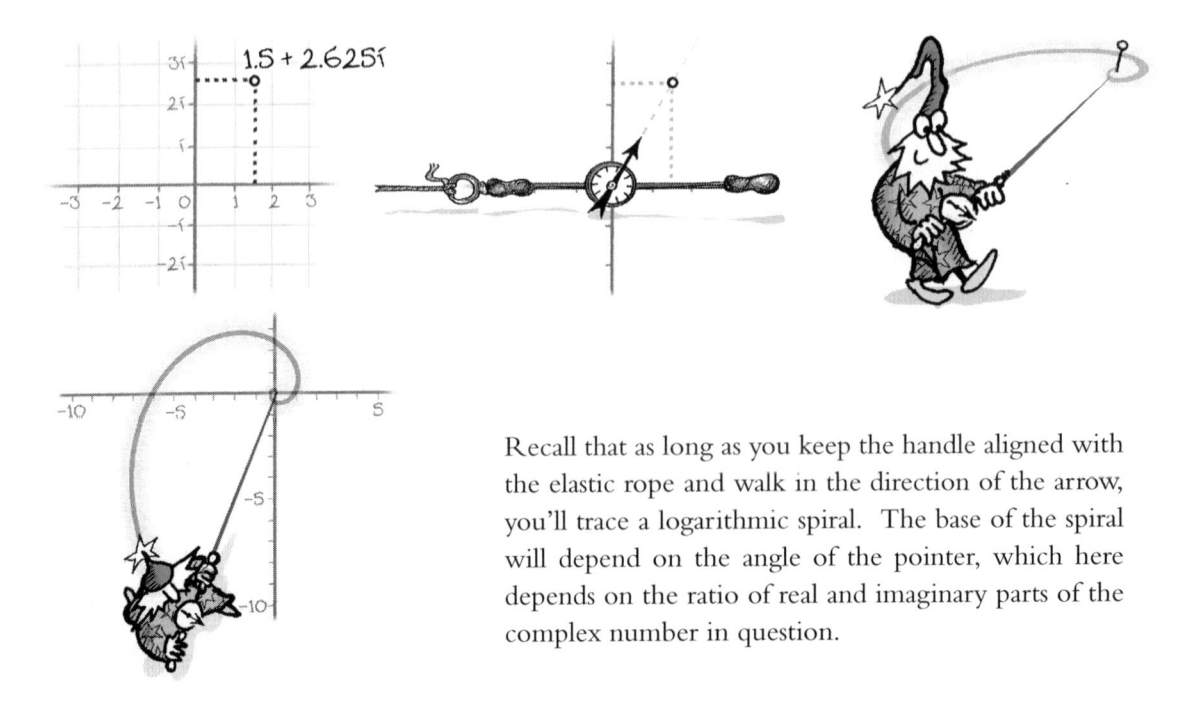

Recall that as long as you keep the handle aligned with the elastic rope and walk in the direction of the arrow, you'll trace a logarithmic spiral. The base of the spiral will depend on the angle of the pointer, which here depends on the ratio of real and imaginary parts of the complex number in question.

Next, we'll mark a sequence of points on our spiral corresponding to the counting numbers 1, 2, 3, 4, 5, ... To do this, for each counting number, we make a circle of that radius and a circle of radius 1, blacking out the inner and outer regions to leave a "visible ring" (as we do when working out logarithms, except that that would usually involve the base-e spiral[4]). Count the number of coils in the visible ring:

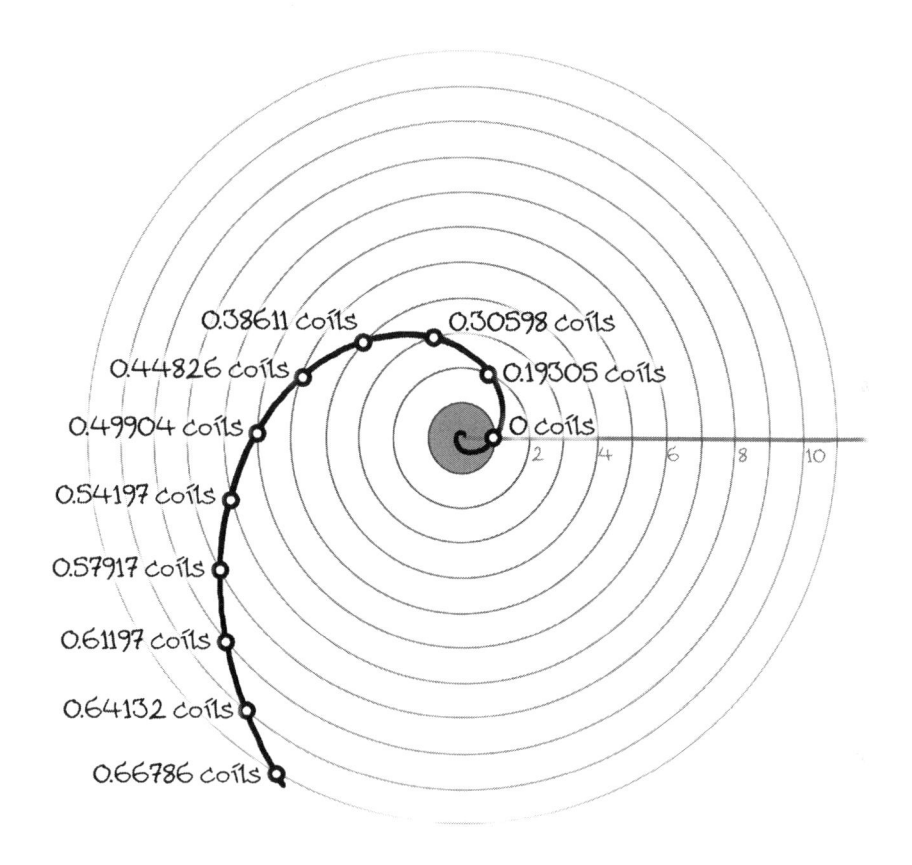

0.38611 coils
0.30598 coils
0.44826 coils
0.19305 coils
0.49904 coils
0 coils
0.54197 coils
0.57917 coils
0.61197 coils
0.64132 coils
0.66786 coils

Only the "inner region" of radius 1 is shown blacked out here. For each radius 2, 3, 4,..., you'll have to imagine the corresponding "outer region" blacked out and the "visible ring" which results.

Now, for each counting number, multiply the number of coils by the real part of your chosen complex number (remember, this is its distance from the imaginary axis) and then travel that many coils clockwise ("inwardly") from 1 towards the centre. So, for the example 3, we had 0.30598... coils. We multiply this by 1.5 (the real part of our chosen complex number) to get 0.45897... and then travel inwardly that many coils (that is, a bit less than half a coil) from 1. Stop there. Now draw an arrow from the centre of the spiral to the point where you've just arrived in the complex plane.

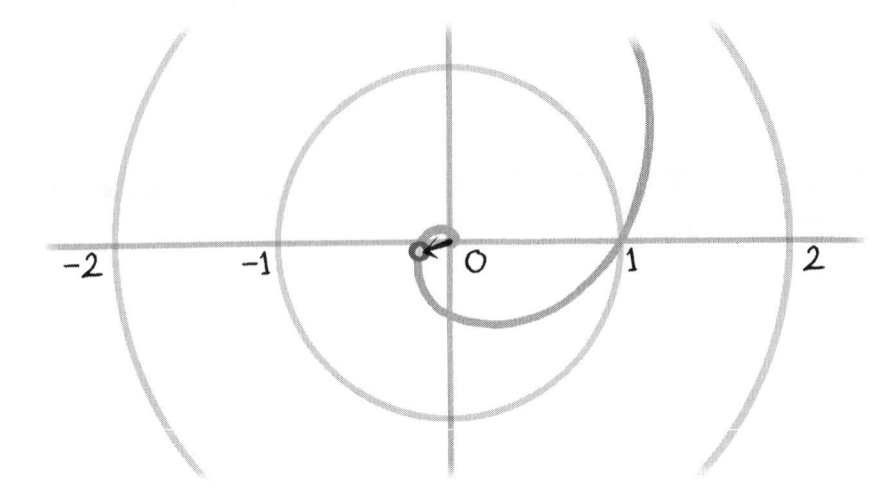

Imagine creating one of these little arrows for each counting number. The arrow for (counting number) 1 will always point to (complex number) 1. Why? Because there are zero coils inside the corresponding "visible ring" (see previous page), so multiplying zero by the real part of your complex number always gives zero and "travelling zero coils" inwardly from 1 just means staying at 1.

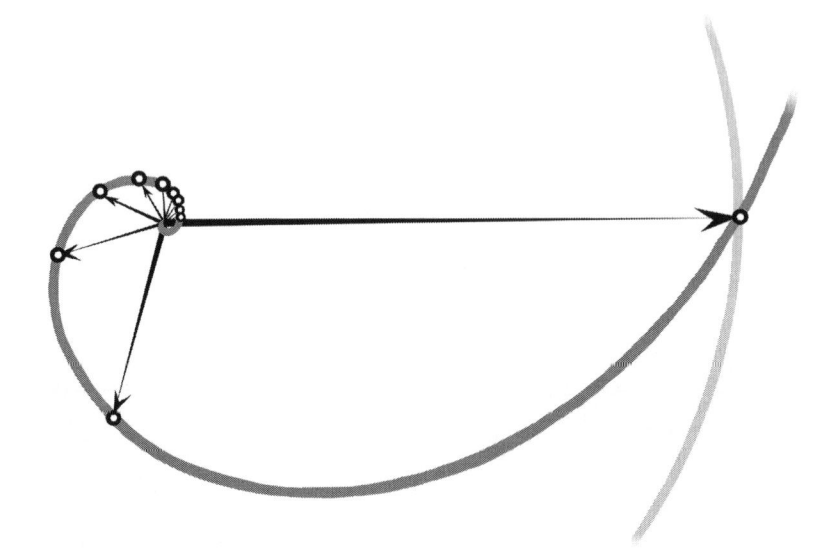

Here we see the arrows that would be constructed for 1, 2, 3, 4, 5, 6, 7, 8,...

Leaving the "1" arrow in place pointing from 0 to 1, imagine successively picking up and moving each of the arrows for 2, 3, 4,... so that they're end-to-end, *making sure you keep them pointing in their original directions*.

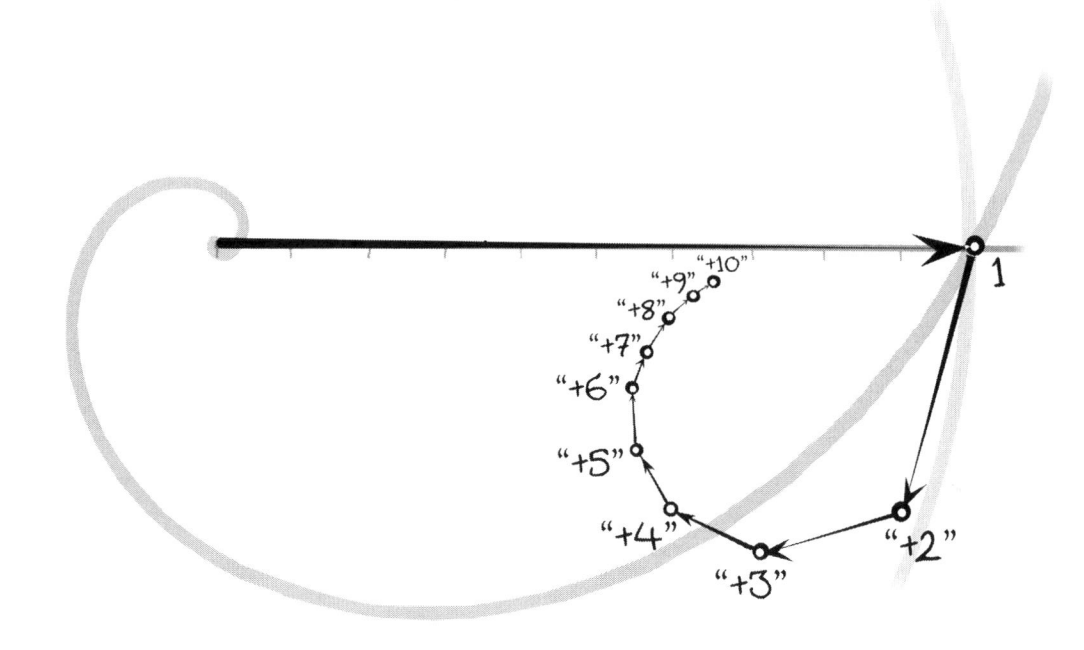

Compare the arrows you see here to those in the previous illustration.

You might recognise this as the process of adding complex numbers together, but in this case we're adding *infinitely many of them*. The little arrows will get shorter and shorter, and *in a way which guarantees that the "chain" they produce will terminate at a specific point*. This is the complex number equivalent of what we saw earlier when we were adding infinitely many real numbers together and getting a finite answer. And as with real numbers, it's usually the case that adding infinitely many (complex) numbers together does *not* give a finite answer (the chain of arrows would just continue on forever, never terminating at any particular point). But in the procedure I've described, there is a guaranteed terminal point.

At the end of this chain of little arrows is the point to which the Riemann zeta function transports the point we originally chose in the complex plane. So, we can imagine that this rather long-winded procedure is the calculation performed by the Riemann zeta sprite before it hits a ball placed somewhere in the relevant half of the complex plane. In the other half of $\mathbb{C}$, this won't work. If you try it, you'll find that your arrows actually get longer rather than shorter as you proceed through the counting numbers, and so joining them end to end produces a chain of arrows which never terminates. But this doesn't mean that the Riemann zeta function is any less "well defined" to the left of that vertical line – it just means that *this procedure* fails to describe its behaviour there. We'll need more subtle mathematics in order to describe what the function does in the left half-plane.

Notice how any other point on the line we drew from 0 through our chosen point would have produced the same spiral, as this only depends on a choice of angle.

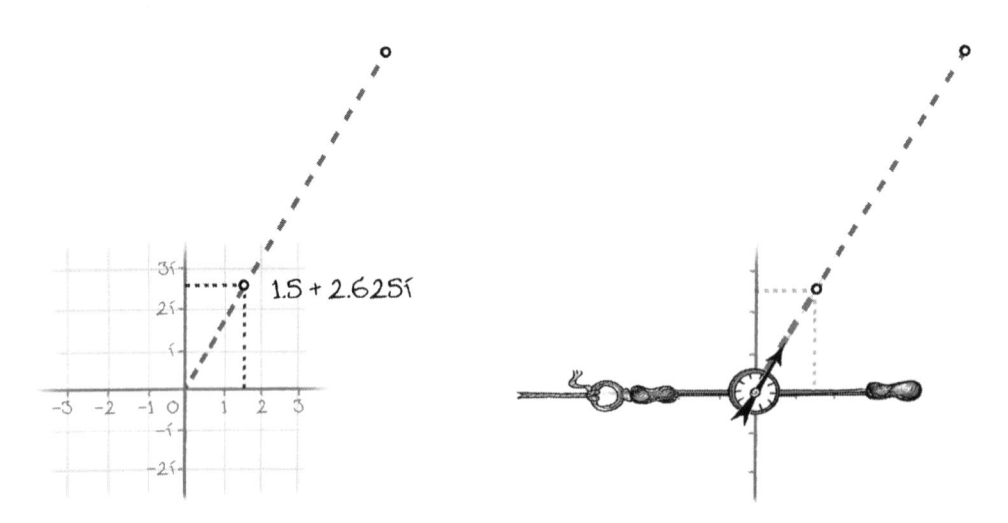

But we'll get a different answer when we apply the Riemann zeta function to such a point because, you'll recall, we used the real part of the chosen complex number in our procedure. Different complex numbers can produce the same spiral, but if they do, then they'll necessarily have different real parts.

If you'd chosen a point on the real axis, then the line drawn through it from 0 would just be a horizontal line, and the equiangular spiral we'd get wouldn't actually be an equiangular spiral as such, it would be half of a line. But, as we saw in Chapter 9, just as a circle can be understood as an extreme (or "degenerate") spiral resulting from the maximum possible choice of angle, this half-line is also an "extreme" spiral, but at the other end of the scale. The circle results from "winding your spiral so tightly that it ceases to be a spiral anymore". This half-line is the result of the minimum possible choice of angle (zero) and produces a spiral so *loosely* wound that it can no longer really be considered a spiral.

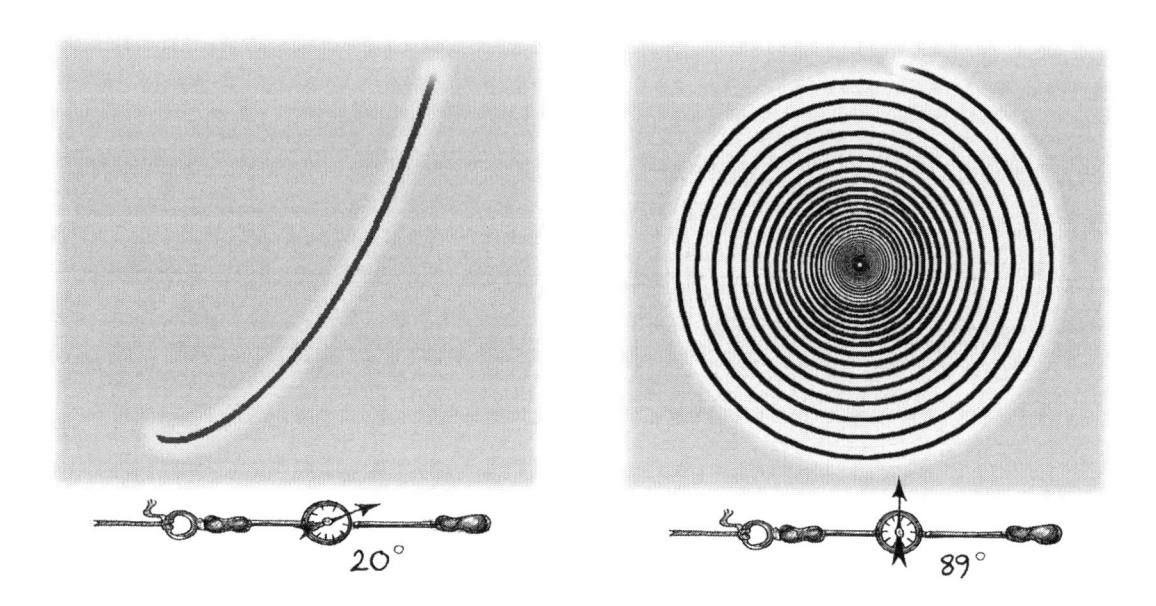

Because we can't "count coils" when our spiral has been flattened out like this, the procedure we just saw cannot be applied. Fortunately, though, we have the Euler zeta function which we can use for points on the real axis, and because the Riemann zeta function is an extension of this, we can be sure that the two methods will produce consistent results near the real axis. That is, if you place two golf balls, one on the real axis and one just slightly off, then the destination of the first can be calculated using the Euler zeta function procedure described earlier, and the destination of the second can be calculated using the Riemann zeta function procedure we've just seen. Carrying these both out, you'd then find the destinations to be appropriately close together, for we know that the Riemann zeta function is *continuous*, which means (loosely) that "nearby points get transported to nearby destinations".

As with the Euler zeta function, the Riemann zeta function is such that you can arrive at exactly the same result using *the sequence of prime numbers and multiplication* rather than *the sequence of counting numbers and addition*. Because of this, the Riemann zeta function further extends our understanding of the relationship between the set of primes and the system of counting numbers.

In the last procedure I described, the arrows generated by the sequence of counting numbers were joined end-to-end to represent the adding together of the associated complex numbers. As with the Euler zeta function procedure seen on pages 40–45, the primes-and-multiplication approach is a bit more complicated than this counting-numbers-and-addition approach.

We choose our point, produce our logarithmic spiral and find the sequence of points on it which are associated with the counting numbers, just as before, but this time we're only interested in the primes, so we discard the arrows associated with 1, 4, 6, 8, 9, 10, 12, 14, 15, 16,..., leaving the familiar sequence 2, 3, 5, 7, 11, 13,...

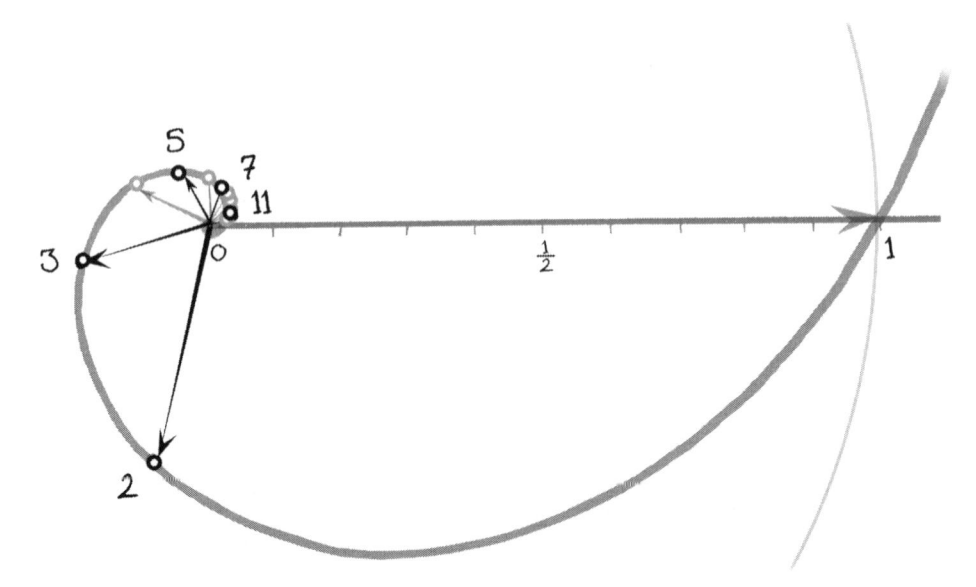

The remaining arrows correspond to complex numbers which we're going to "combine multiplicatively", although it's not as simple as just multiplying them together. Don't worry too much if this seems confusing to you – we're not actually going to use it.

The main point here is that we can get the same result using either the counting-numbers-and-addition approach or the primes-and-multiplication approach.

For each complex number represented by an arrow (there's one for each prime), we do the following.

First, we subtract it from the complex number 1. We've seen a general method for subtracting complex numbers, but one simple way to carry out this particular "1 minus..." subtraction is to take an arrow to the point in question, reflect it across 0 and then join the resulting arrow end-to-end with the "1 arrow":

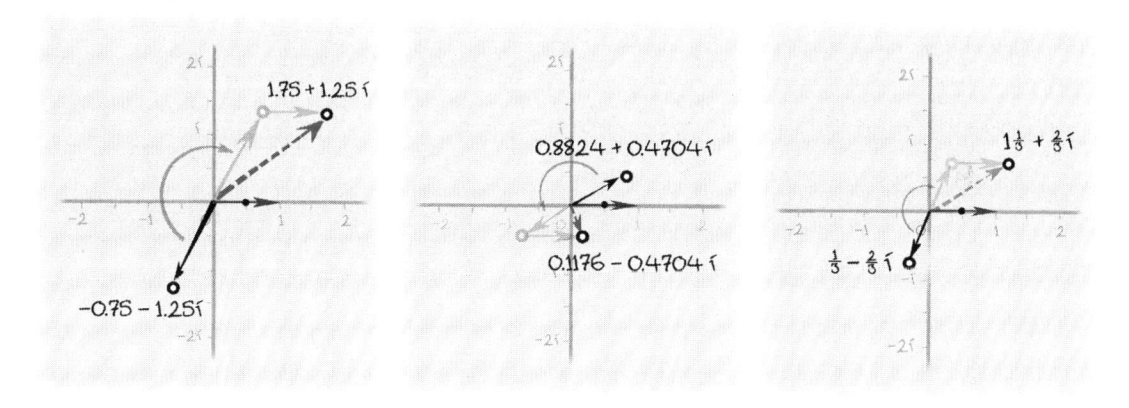

We then have to divide 1 by the resulting complex number. To do this geometrically, we first take an arrow from 0 to the point and reflect it across the real axis:

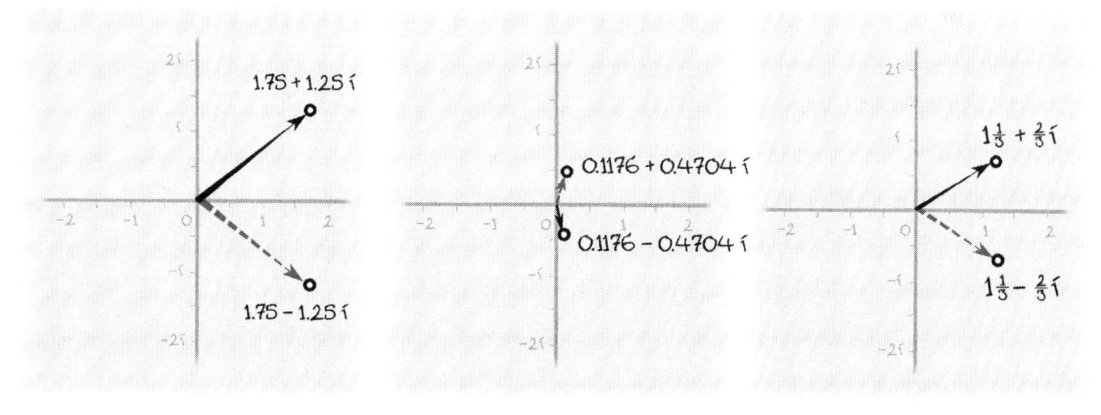

We then need to change the length of the arrow in the following way. Stretch or shrink it (whichever is appropriate) so that it ends up with length 1...

63

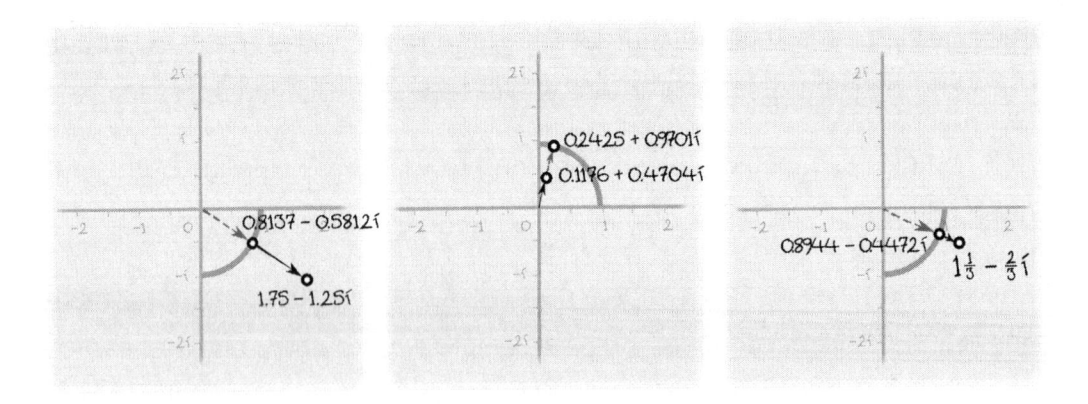

...then perform the same magnitude of stretching or shrinking once again:

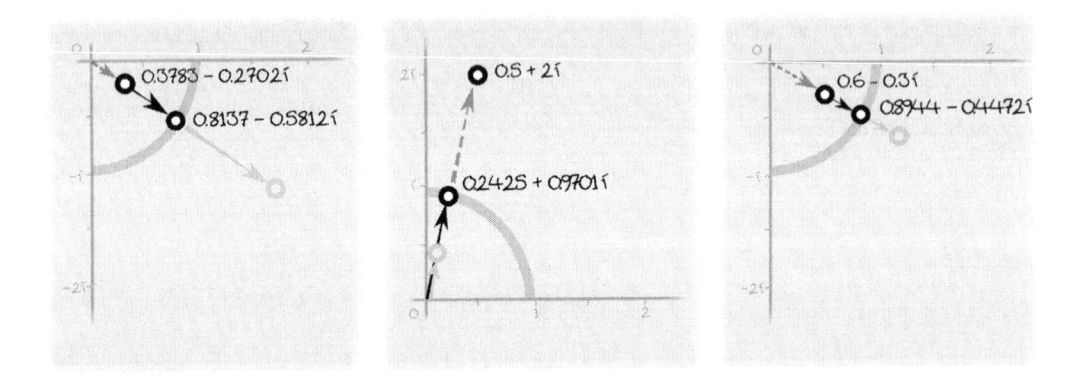

Having done all of this for each prime-number-related arrow, we have an infinite sequence of these flipped, shifted, flipped again and doubly-stretched-or-shrunk arrows. *Now* we multiply them all together. We saw how to multiply two complex numbers back on page 20 (it involved elastic triangles, remember?). So we start with the first two of our arrows (complex numbers) and multiply them, giving a new one. We then multiply that with the next arrow, multiply the resulting arrow by the next, and again, and again... This is the complex number equivalent of multiplying infinitely many numbers together and, as with the real number version we saw earlier, there are situations where such infinite multiplication can yield a finite result. In this case, as long as we start with a point in the right half-plane, it can be proved that the result of the procedure we've just seen, culminating in the multiplication of infinitely many complex numbers, *must* produce a (finite) complex number.

Phew! Fortunately, you won't need to retain *any* of that, apart from the all-important fact that the complex number which this primes-and-multiplication procedure produces is the same one we got before using the much simpler counting-numbers-and-addition procedure. Clearly, the addition procedure is much easier to work with, but the key here is that *it can be replicated with the primes and multiplication* – this is what's mathematically interesting. It tells us something very important about how the sets of prime numbers and counting numbers interrelate (and therefore about how addition and multiplication interrelate). I made the same claim earlier when we saw, analogously, how the Euler zeta function's behaviour could be equivalently calculated using either counting-numbers-and-addition or primes-and-multiplication. But in the extended context of $\mathbb{C}$, we're able to use this property to learn a *lot* more about the number system.

To conclude this chapter, we'll quickly review some of the important properties of the Riemann zeta function:

✫ It's defined on the whole of the complex plane except at the point 1.

✫ It extends the Euler zeta function. That is, when we choose a point on that part of the real axis to the right of 1 (where Euler's function is defined), the two functions "transport it to the same destination" (which will also be on the real axis).

✫ It's "deterministic". This is true of *every* function, but still worth stressing: if we pick the same point several times (imagine placing a golf ball there), it gets transported to exactly the same destination each time.

✫ Unlike the Euler zeta function, it *does* have zeros (points in $\mathbb{C}$ which it transports to 0). We've already found such points at $-2, -4, -6, -8, \ldots$ (the negative even integers), but there may turn out to be others – we shall see.

✫ It can sometimes transport two different points to the same location (we've just noted that both −2 and −4 get transported to 0, for example).

✫ It's *continuous*, which means (roughly) that "nearby points get transported to nearby destinations".

✫ The sole point in $\mathbb{C}$ where it's not defined, that is, 1, is called the *pole* of the Riemann zeta function (some complex functions have more than one pole, some don't have any – the Riemann zeta function has just this one). Points chosen ever closer to 1 get transported farther and farther out onto $\mathbb{C}$, in such a way that you *could* think of the point 1 as being transported "out to infinity" by the zeta function (although it's not clear exactly what that would mean).

✫ It transports the point 0 to the point −1/2.

✫ It transports the point 2 to 1.64493..., which is what you get when you multiply π by itself and then divide by 6 (we saw this for the Euler zeta function, which "coincides with" Riemann's zeta function at such points). "1 in 1.64493..." happens to be the probability of two randomly chosen counting numbers being "relatively prime" (their factorisations having no prime factors in common).

chapter 19
the Riemann zeta zeros

Mathematicians have developed some tremendously complicated methods to explore the behaviour of particular functions, or whole families of functions. This stuff is intellectually challenging in the extreme, so we shall have to be content to potter about on the complex plane golf course with our Riemann zeta function sprite and make a few observations. The mathematical truths which I'll state will have to be accepted with the understanding that rigorous mathematical arguments *have* been constructed and have survived the most severe scrutiny imaginable in order to establish these facts. And as far as "advanced human knowledge" goes, mathematical proofs are as unanimously agreed upon as anything gets.

As I've already mentioned, when mathematicians study a function (real or complex), among the first things they're likely to consider are its zeros, a "zero" of a function being any number (a real number on the usual number line, or a complex number in the complex plane) which the function transports to 0.

As a way of picturing the search for the zeros of the Riemann zeta function, we've introduced a single hole to our complex plane golf course, at the point 0, the only other feature of the infinite, flat plane being the pole at 1.

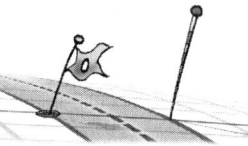

Recall that the aim of our game is to find locations on the course where, if a ball is placed there, the Riemann zeta function sprite will hit it into the hole. If you place your ball at a point, summon the sprite to hit it and then witness it arcing beautifully into the 0-hole, then you've just found a zero of the Riemann zeta function. These complex numbers will be referred to variously as *Riemann zeta zeros*, *Riemann zeros* or just *zeta zeros* (these all being short for "zeros of the Riemann zeta function"). In the previous chapter, we found a whole sequence of zeta zeros, at the negative even integers −2, −4, −6, −8, ... The question that remains is whether there are any others.

As the complex plane (being an abstract mathematical domain) could be thought of as having a kind of "timelessness", we'll imagine that we have all of eternity to experiment with the Riemann zeta function via its (tireless and eternal) golfing sprite.

One thing we can make use of is the fact that the zeta function is continuous. Because this means that balls placed close together get hit to nearby destinations, we can use a strategy rather like the one used in the infant's game of seeking some unknown object, where clues are given in the form of "warmer" (meaning "you're getting closer") and "colder" (meaning "you're getting farther away").

Through an eternity of trial-and-error, we would eventually convince ourselves that there are no Riemann zeros to the right of the vertical line through the pole at 1, and that, apart from those which we found at −2, −4, −6, −8, ..., there are none to the left of the vertical line through the hole at 0. As I've already stated, these are well-established mathematical facts. Riemann was able to prove them almost as soon as he'd extended Euler's zeta function to $\mathbb{C}$, but we needn't be concerned with the details of how he did this. If you really want to know (and have some grasp of mathematical notation and reasoning), you can find out more in Appendix 11.

The zeros at −2, −4, −6, −8, ... are very well understood by mathematicians, their entirely unmysterious nature having led to them being called *trivial zeros* of the Riemann zeta function. Any other zeros, which we now know must lie in the vertical strip between 0 (the hole) and 1 (the pole), will be called *nontrivial zeros*. This vertical strip has come to be known as the *critical strip* of the Riemann zeta function.

So we've literally narrowed down our search to one little strip within $\mathbb{C}$. This makes things easier. Through a further eternity of trial-and-error golf sprite experimentation, we should start to "home in" on the first of our nontrivial zeros, which lies exactly halfway across the strip (that is, on the vertical line through 1/2) and just over 14 units up from the real axis.

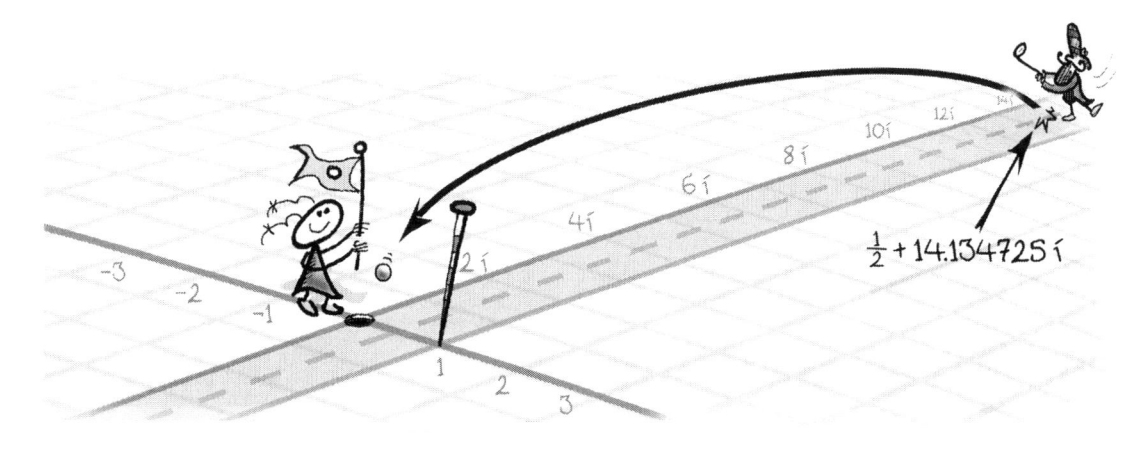

The location of this zero, we'd find, would be around 1/2 + 14.134725*i* (here, the 1/2 is precise, while the 14.134725 is approximate – does this particular number seem familiar to you from anywhere?).

Seeking nontrivial zeros in the critical strip *below* the real axis, we'd similarly find a nontrivial zero at about 1/2 − 14.134725*i*.

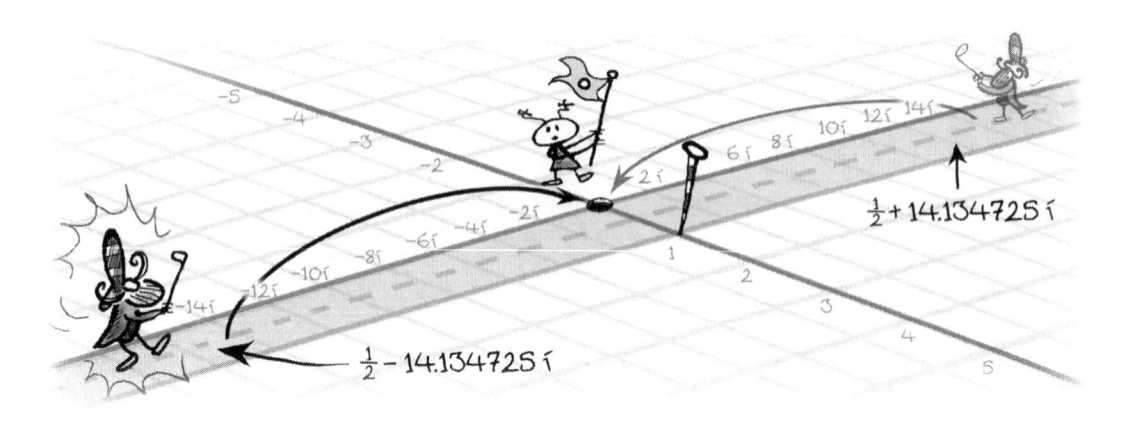

In this image and others which follow, the vertical direction has been rescaled for ease of illustration.

Working our way up the upper half of the strip, we'd find the next zero at about 1/2 + 21.02204*i*, and working our way down the lower half, we'd find another one at about 1/2 − 21.02204*i*.

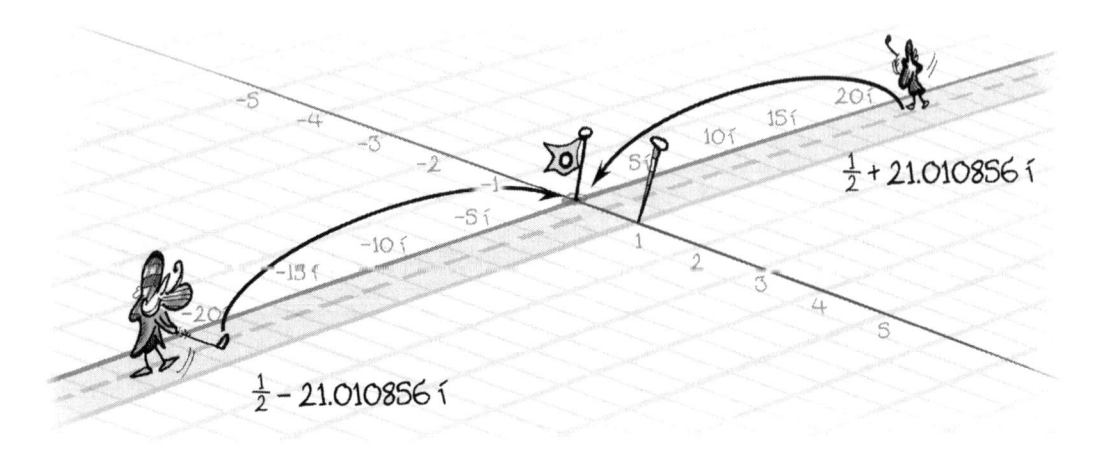

This suggests that the zeros come in pairs: "mirror image" pairs which are reflections across the real axis. This is indeed the case: Bernhard Riemann was able to prove it early on in his investigation of the zeta function, using the theory of functions on the complex plane which was available in the 1850s. If there's a zero above the real axis, then it will always be matched by a zero below the real axis, its mirror image.

The next few pairs of nontrivial zeta zeros (as we move away from the horizontal axis) would be found to be at (approximately)

$$1/2 + 25.010858i \quad \text{and} \quad 1/2 - 25.010858i,$$
$$1/2 + 30.424876i \quad \text{and} \quad 1/2 - 30.424876i,$$
$$1/2 + 32.935062i \quad \text{and} \quad 1/2 - 32.935062i,$$
$$1/2 + 37.586178i \quad \text{and} \quad 1/2 - 37.586178i,$$
$$1/2 + 40.918720i \quad \text{and} \quad 1/2 - 40.918720i.$$

Again, the imaginary parts of these numbers may seem somehow familiar to you (or you might even remember *exactly* where you've seen them before).

The pattern we see emerging here is that all of the nontrivial zeros seem to have real part equal to 1/2: the seven mirror image pairs which we've seen so far all lie on the vertical line through 1/2. This vertical "axis" running up the middle of the critical strip has become known as the *critical line*.

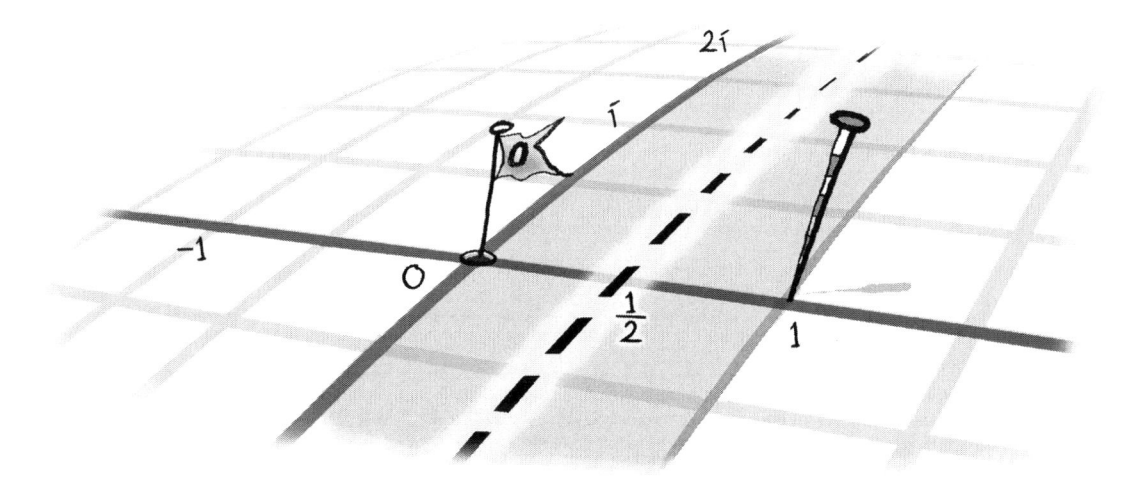

Because of (a) this seemingly significant status of the critical line and (b) the mirror image reflection of nontrivial zeta zeros across the real axis, the point 1/2, which is the midpoint between the hole (0) and the pole (1), appears to be acting as some kind of important "pivot" for the behaviour of the zeta function.

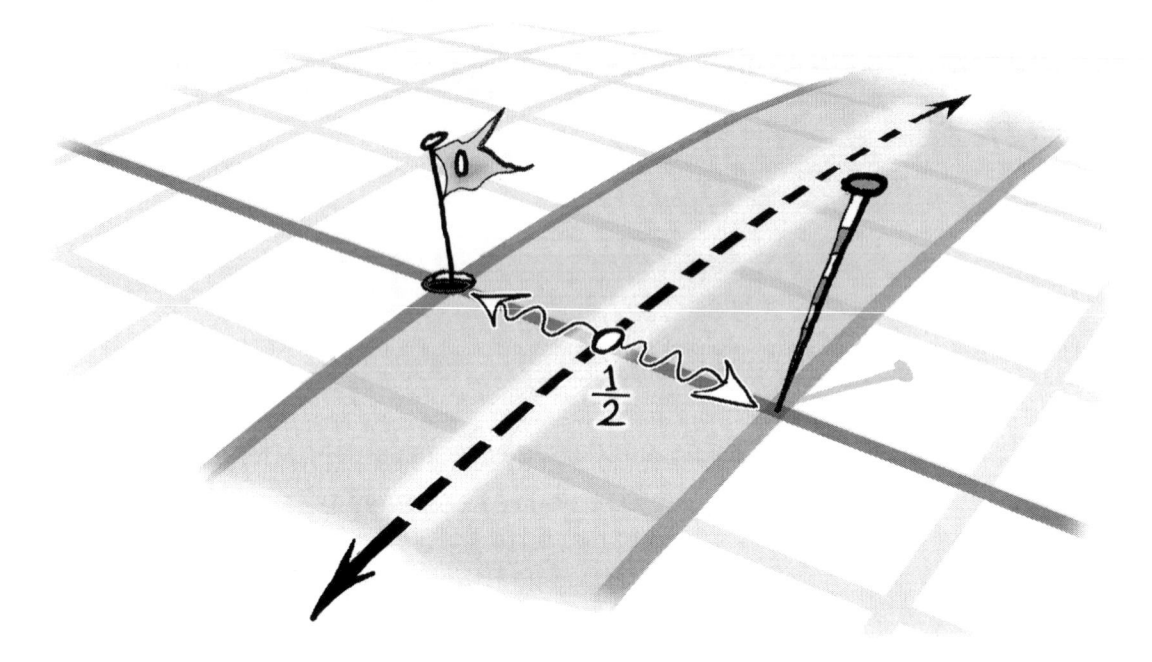

In fact, there's a result known as the *functional equation* of the Riemann zeta function (described by Princeton mathematician Peter Sarnak as "*perhaps the most important thing we know about the zeta function, as far as its structure...*"[1]) which relates the behaviour of the function at pairs of points "reflected across" the point 1/2.

What does this "relates the behaviour" mean?

First, we'll need this:

It's not too difficult to show that if you take a complex number and subtract it from 1, then the resulting complex number is represented by a point in $\mathbb{C}$ which is the same distance from the point 1/2, but in the opposite direction. Here are some examples (it may help to refer back to page 63):

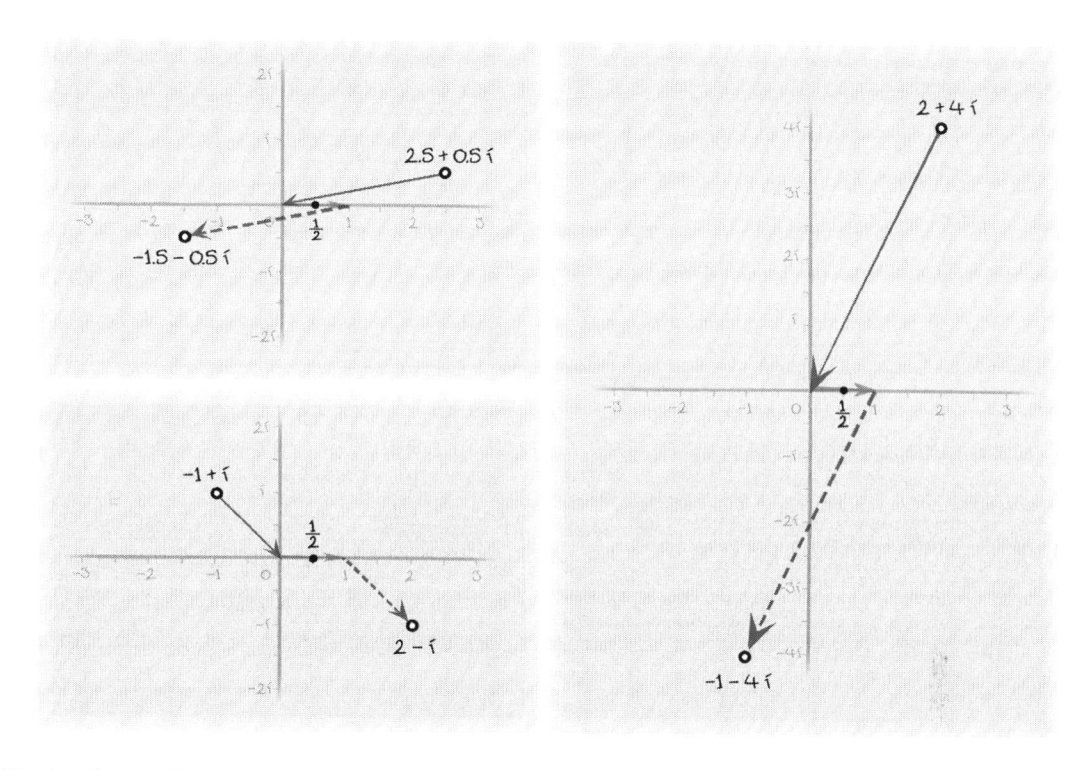

Notice how the hole and the pole are also mutual reflections across the point 1/2. This corresponds to the simple fact that $1 - 0 = 0$.

Now, imagine placing a golf ball at a point on $\mathbb{C}$ and another at its reflection across 1/2, as with any of the pairs just illustrated (only the pair 0, 1 is disallowed). We summon the function sprite and the two balls get hit to a pair of destinations (points representing complex numbers). The functional equation provides a mathematical relationship between these two complex numbers. It looks like this:

$$\frac{\Gamma(\frac{s}{2})\zeta(s)}{\pi^{\frac{s}{2}}} = \frac{\Gamma(\frac{1-s}{2})\zeta(1-s)}{\pi^{\frac{1-s}{2}}}$$

We're not going to worry too much about the details (you can have a look at Appendix 12 if you're brave). The key observation here is that on one side of the equation, we have something involving "s", and on the other side, we have the same thing, but involving "1−s". The "s" represents any choice of complex number, and "1−s" is the complex number you get when you subtract it from 1, which, as we've seen, is the point in $\mathbb{C}$ you get by reflecting the point associated with "s" across the point 1/2. The behaviour of the Riemann zeta function at this pair of points is being shown to be related.

Using this equation, it's possible to deduce what the zeta function will do on one side of the point 1/2 by using knowledge of what it does on the other side. Our original definition of the function (remember those intricate procedures involving spirals in the last chapter?) was restricted to the half-plane to the right of the critical strip. Using the functional equation, we can deduce what the zeta function does at any point in the half-plane to the *left* of the critical strip, because any such point has a reflection across 1/2 which lies in the half-plane where we *know* what the function does. To work out what the zeta function does *within* the critical strip requires a further trick, which (if you have sufficient mathematical confidence) you can read about in Appendix 13.

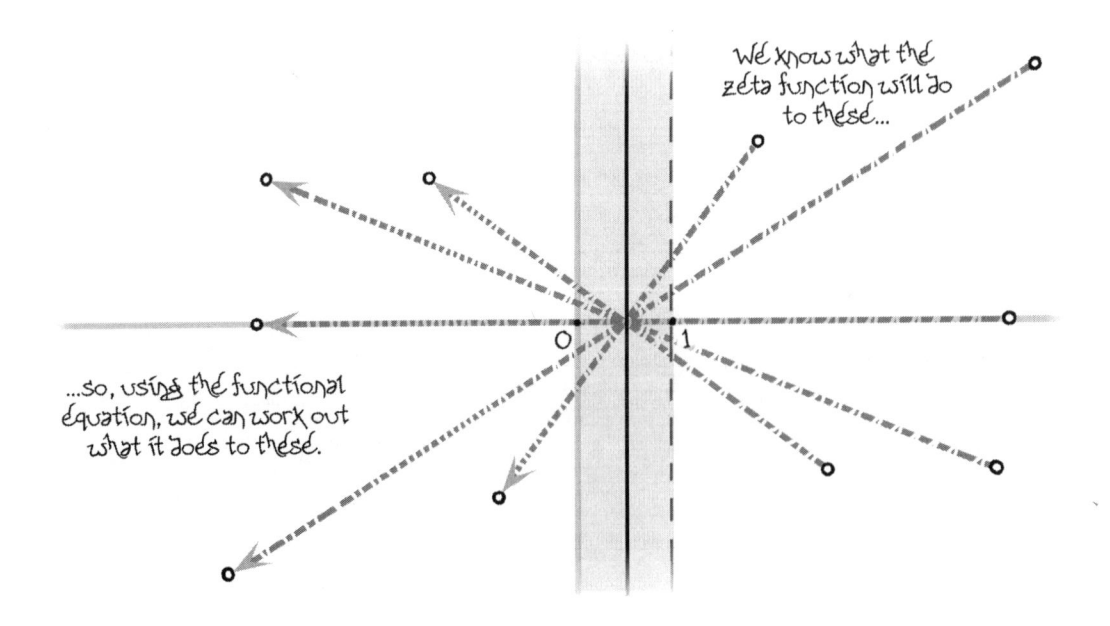

We know what the zeta function will do to these...

...so, using the functional equation, we can work out what it does to these.

74

Note that the functional equation *doesn't* say that the zeta function transports these pairs of "reflected" points to the same destination. Rather, it provides a mathematical relationship between their destinations. However, this relationship does have the helpful property that *if one of the points gets transported to zero, then the other must also*. In other words, if you find a zero of the zeta function, then its reflection across the point 1/2 must also be a zero.

So what then of the trivial zeros at −2, −4, −6, −8,...? These are zeros of the zeta function, and they reflect across the point 1/2 to 3, 5, 7, 9,..., respectively (all points on the positive real axis).

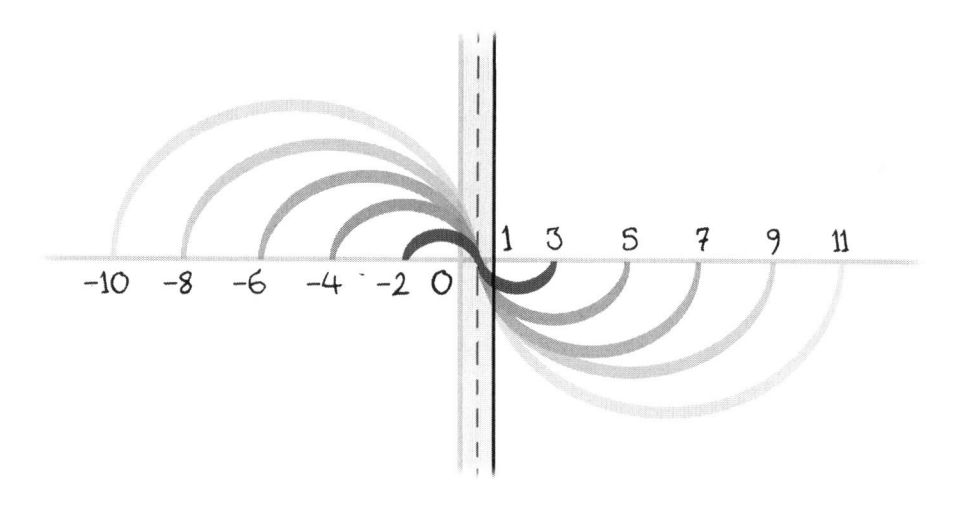

So shouldn't these also be zeros? No. The points −2, −4, −6, −8,... get treated differently by the functional equation. You can see that it involves "$\Gamma(s/2)$". The "Γ" is another Greek letter, a capital "gamma", and it denotes another special function called the *gamma function*[2]. The gamma function is defined everywhere on $\mathbb{C}$ *except at the negative integers*. Because of the way it's defined, *all* negative integers are poles of the gamma function. "$\Gamma(s/2)$" literally means "the point in $\mathbb{C}$ where you end up if you take your chosen complex number, divide it by 2, and then apply the gamma function". Dividing any of −2, −4, −6, −8,... by 2 gives a negative integer, so the gamma function cannot be applied. For this reason, the functional equation works everywhere in $\mathbb{C}$ *except* at −2, −4, −6, −8,...

75

All the nontrivial zeros we've found thus far (the ones inside the critical strip) have come in pairs, reflected across the real axis. Because all of them have also been found to lie on the critical line (the vertical line through the point 1/2), you might think that this reflection property is just a simple consequence of the functional equation. But there's more to it than that. Any zero which was found *off* the critical line would automatically give rise to two more zeros, one reflected across the real axis (directly below the original zero) and one reflected across the point 1/2, as shown in the left-hand illustration below:

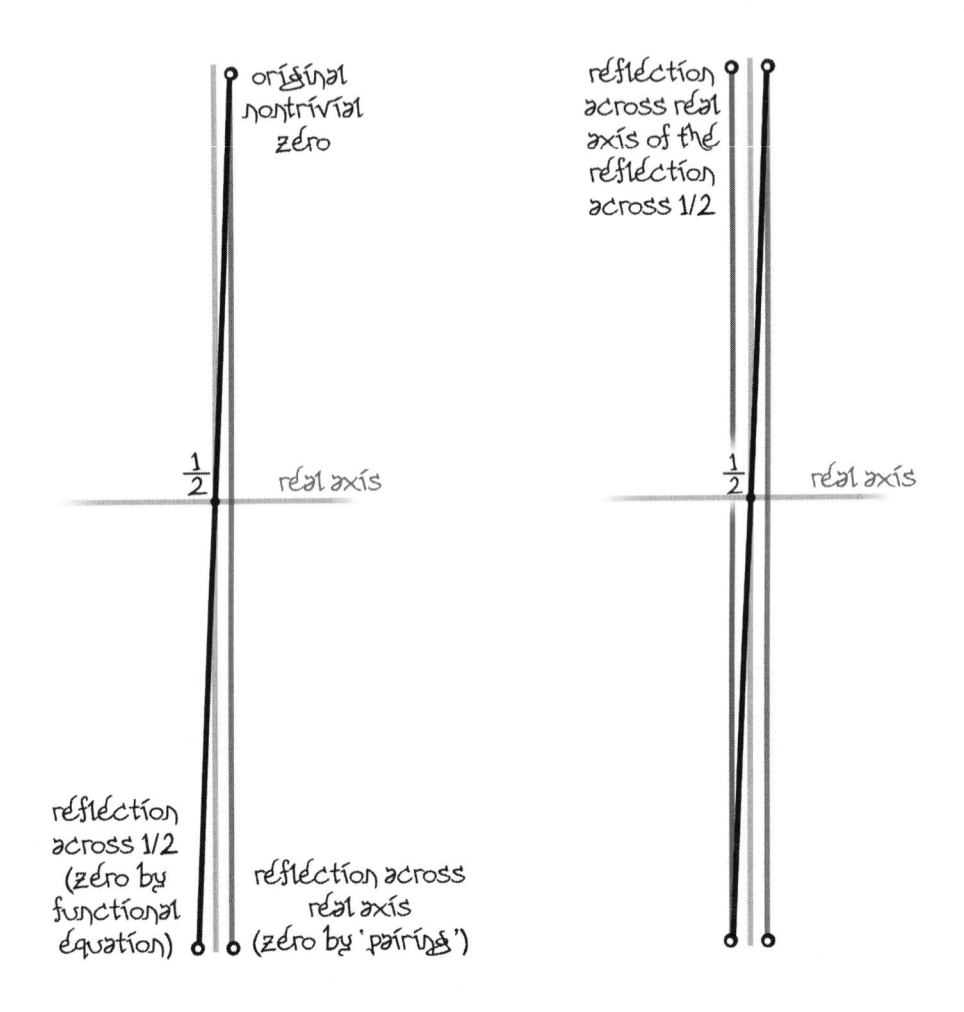

The second of these would then produce *another* nontrivial zero — *its* reflection across the real axis, as shown in the right-hand illustration above.

So, if a nontrivial zero lies *on* the critical line, we automatically get a *pair* of nontrivial zeros, and if it lies *off* the critical line, we get four nontrivial zeros:

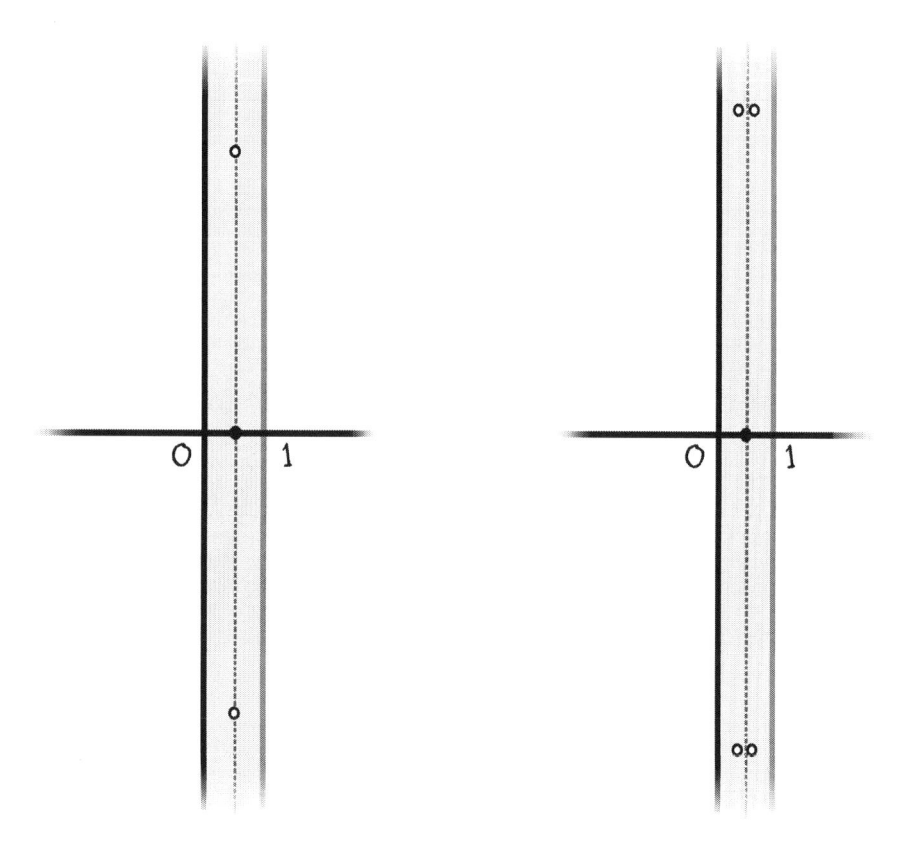

Incidentally, despite not having an infinite amount of time (he died at age 39) or the power to summon golfing sprites, Riemann was able to compute the locations of the first few pairs of zeta zeros. This was quite a remarkable feat, considering what he had available to him – these days, they're found by powerful computers (yes, new zeta zeros are still being found). The zeros he found were all on the critical line and he appears to have been quite confident that *all* nontrivial zeros would be on this line. This suggestion appears in his historic 1859 paper, and although he was unable to find a way to prove it, he doesn't seem to have believed that it was a particularly difficult or noteworthy problem. Riemann's comments suggest that he imagined someone else *would* prove it before too long[3], but at the time of my writing this (2011), no one yet has[4].

Riemann *was* able to prove a number of other things about the zeros of his zeta function, though:

✯ Apart from the trivial ones at −2, −4, −6, ..., they all lie in the critical strip[5].

✯ They come in "mirror image pairs", reflected both across the real axis and across the point 1/2 (we've already seen this – notice that if they *do* all lie on the critical line, then these two facts amount to the same thing).

✯ There are infinitely many of them.

✯ They tend to cluster together as we travel up the critical strip and, statistically, this clustering is governed by a law involving a logarithm[6].

Riemann's 1859 paper was called (translated into English) "On the Number of Primes Less Than a Given Magnitude". Recall that the Prime Number Theorem (PNT), which gives a logarithmic law describing the statistical thinning out of the primes along the number line, had been proposed as early as the 1790s, but wasn't proved until 1896. So this result was known but unproved for the duration of Riemann's life. Although he was unable to prove the PNT (his fundamental aim), his paper linked it to a certain fact about the locations of nontrivial zeta zeros. The proofs of the PNT which Jacques Hadamard and Charles de la Vallée Poussin published in 1896 simply involved the proof of this fact. So their proofs were built on the foundation of Riemann's work on the zeta function.

It's worth re-emphasising that the point 1/2 has the property of being halfway between 0 (the hole) and 1 (the pole). You might remember that it was these two key points which we started off marking on an unmarked line in Chapter 16, after which all of $\mathbb{N}$, $\mathbb{Z}$, $\mathbb{Q}$, $\mathbb{R}$ and $\mathbb{C}$ sprung into life. 0 is known to mathematicians as the *additive identity* (the unique number you can add to anything without changing it) and 1 is similarly known as the *multiplicative identity* (the unique number you can multiply anything by without changing it). I've already hinted that the behaviour of the Riemann zeta function is deeply connected to the "difficult" relationship between addition and multiplication. So, the fact that the midpoint between the additive and multiplicative identities should act as a "pivotal" point in the zeta function's behaviour seems appropriate.

We've seen that the "awkward" distribution of primes is in some sense the result of this uneasy relationship between addition and multiplication. The Riemann zeta function, with its twin definitions (as seen in Chapter 18), can be seen to be intimately linked to the way the sequence of primes relates to the system of counting numbers and the way that multiplication relates to addition.

The fact that the PNT can be shown to be equivalent to a fact about the locations of the zeros of the zeta function is in keeping with these observations since it concerns the overall behaviour of the primes and how they distribute in $\mathbb{N}$. As we'll soon see, subtle aspects of the primes' collective behaviour can be related to questions concerning the locations of the nontrivial Riemann zeta zeros, so the PNT can naturally be linked to such issues.

So, we begin to see how the nontrivial zeta zeros and the prime numbers are interrelated. But don't make the mistake of assuming that there's some kind of correspondence *between individual primes and individual zeta zeros*. This is *not* the case – the relationship is far more interesting than that.

The sudden "blossoming" of Euler's zeta function which occurred once Riemann had extended it to the complex plane was a major event in the history of human thought (although only the tiniest proportion of humanity is aware of this yet), suddenly revealing that the relationship between the primes and the counting numbers is even more of a mystery than previously thought.

Chapter 20
Zeta Zeros and Spiral Waves

Back in Chapter 10, we saw that rather than following a pattern in any familiar sense, the seemingly random distribution of prime numbers has a tendency to thin out at an "average" rate (rather like the edge of a cloud might). But the actual locations of the primes generally *deviate* from this tendency to thin out. In any given stretch of the number line they might squash together or spread out far more than their "average tendency" would suggest they should.

In Chapter 14, we saw how this deviation from the average tendency can be described precisely in terms of an infinite number of what I called *spiral waves*. These waves, in our rather fanciful visualisation, were shown as being related to the lengths of the trumpets being blown by an infinite host of (paired) angels.

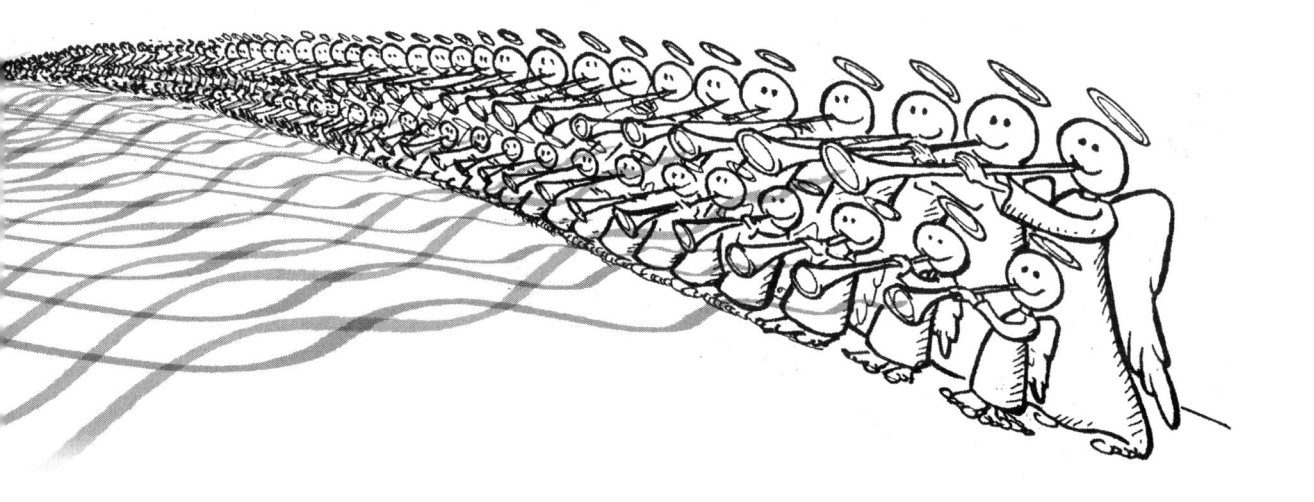

These spiral waves, in particular what I've called their "specifications" (the numbers which exactly describe them – something like amplitude, frequency, *etc.* for ordinary sine waves), were left as a mystery at that point. We're now in a position to gain a better understanding of where these come from.

You might have noticed that the numbers 14.134725..., 21.022040..., *etc.* which appeared in the previous chapter as the heights (or imaginary parts) of the first few nontrivial zeros in the upper half of the critical strip also featured back in Chapter 15 as the "frequencies" of the first few spiral wave pairs.

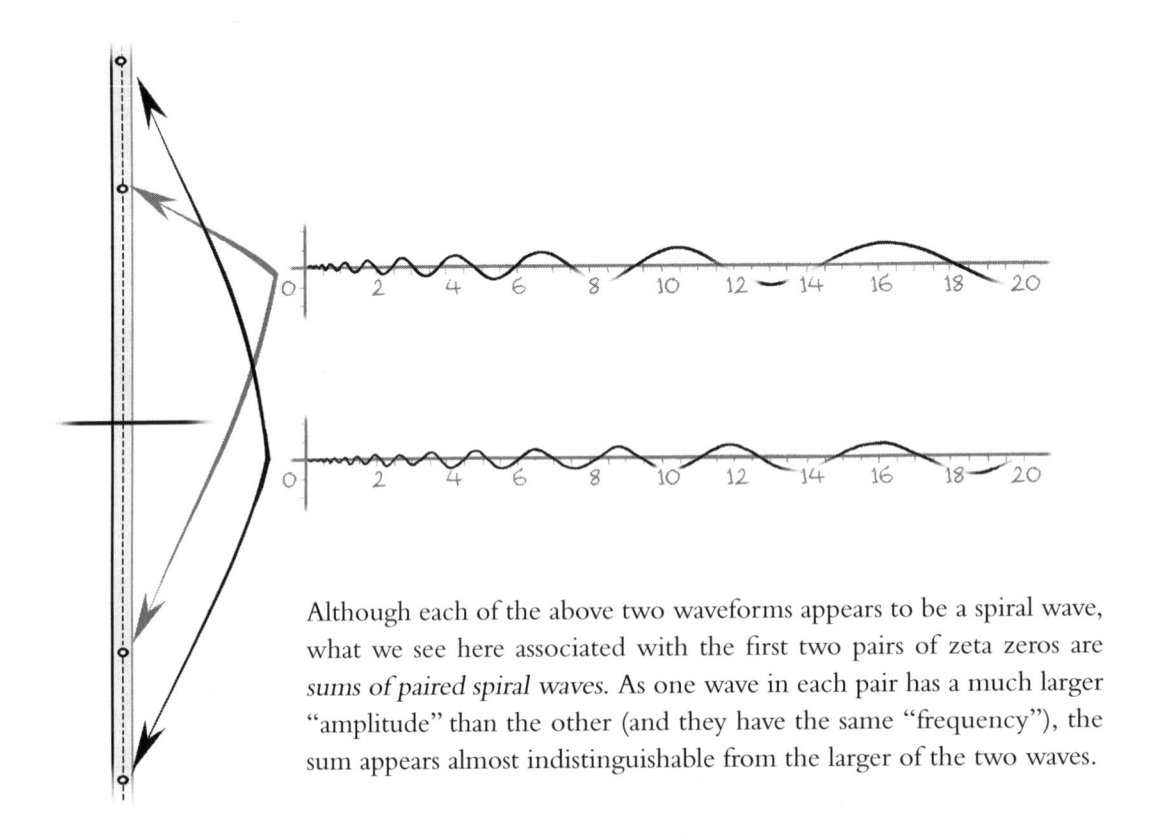

Although each of the above two waveforms appears to be a spiral wave, what we see here associated with the first two pairs of zeta zeros are *sums of paired spiral waves*. As one wave in each pair has a much larger "amplitude" than the other (and they have the same "frequency"), the sum appears almost indistinguishable from the larger of the two waves.

This was not a coincidence. Bernhard Riemann's achievement went far beyond laying the foundations for the proof of the PNT. His exploration of the zeta function led him to develop what's become known as the *explicit formula*, which shows exactly how the "primeness count staircase" function we met in Chapter 12...

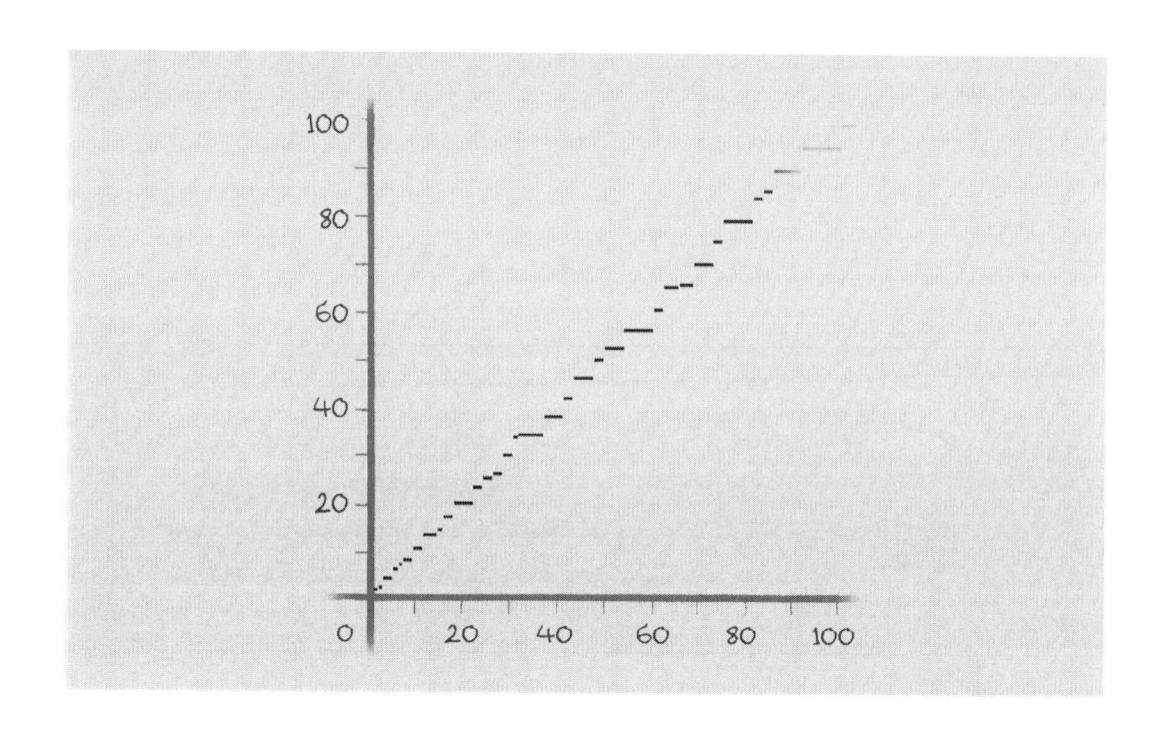

...can be built up from an "average behaviour" plus a couple of extra bits...

...with an infinite sequence of paired spiral waves added in to produce the fine detail:

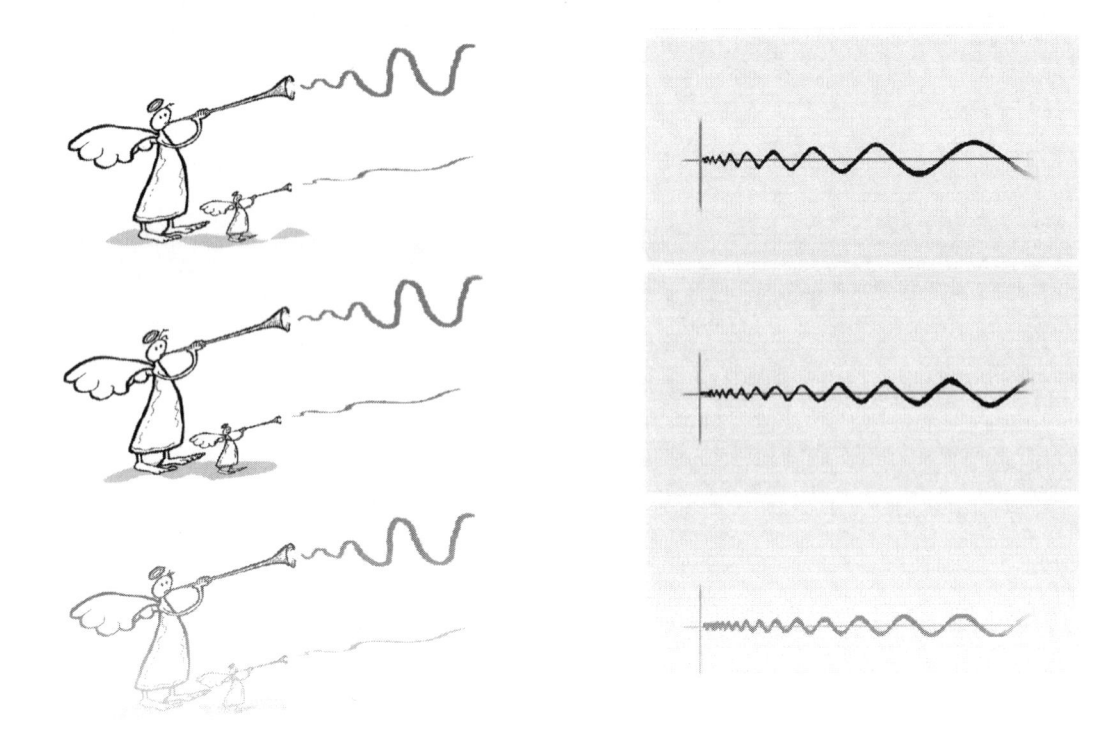

We saw in Chapters 14 and 15 why the fact that you can build the "primeness count staircase" in this way is equivalent to the fact that the deviation...

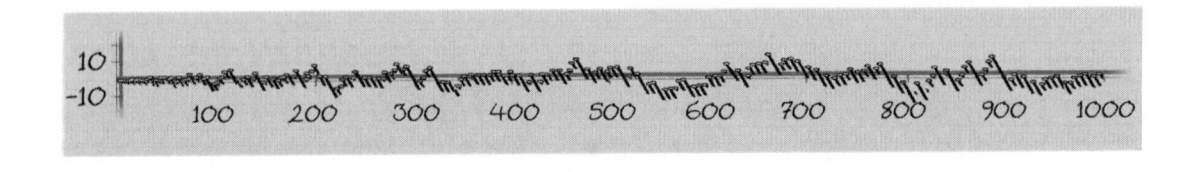

...can be built as the sum of the same spiral waves plus the same two extra bits.

The important thing about Riemann's explicit formula is that *it directly relates the prime numbers to the nontrivial zeta zeros*. If you knew *all* the locations of the nontrivial zeros (which you can't, of course, there being infinitely many), then you'd be able to use this formula to find the exact locations of *all* the primes. There are, in fact, many different versions of the explicit formula, all saying slightly different versions of the same thing, some of them developed quite a bit later than Riemann's original. The explicit formula we're interested in looks like this[1]:

$$\psi(x) = x - \log(2\pi) - \tfrac{1}{2}\log\left(1 - \tfrac{1}{x^2}\right) - \sum_{\rho} \frac{x^\rho}{\rho}$$

As usual, you won't need to understand this. We'll just have a quick look. The thing on the left-hand side represents the logarithmic prime counting function, as illustrated by the staircase graph we saw a couple of pages back. The "x" at the beginning of the right-hand side represents the "average behaviour" of this prime counting function, as illustrated by the diagonal line graph. The "$-\log(2\pi)$" and the "$-\tfrac{1}{2}\log(1-1/x^2)$" represent the two "extra bits", and the final item of notation to the far right represents the sum of all the spiral waves, where the "ρ" (Greek letter *rho*, not "p") stands for a typical nontrivial zero of the Riemann zeta function. We're looking at a sum which involves all (infinitely many) nontrivial zeros, ultimately a sum of infinitely many spiral waves where the "frequencies" of these waves correspond to the heights (imaginary parts) of the nontrivial zeros.

It's worth pointing out that in Riemann's calculations which led to this formula, each of the four parts of the right-hand side can be traced back to "significant points" for the zeta function in $\mathbb{C}$: the "x" can be seen to come from its pole at 1, the "$-\log(2\pi)$" can be linked to 0, the "$-\tfrac{1}{2}\log(1-1/x^2)$" can be related back to the sequence of trivial zeros ($-2, -4, -6, -8,\ldots$) and, as just explained, the final part originates with the set of all nontrivial zeros.

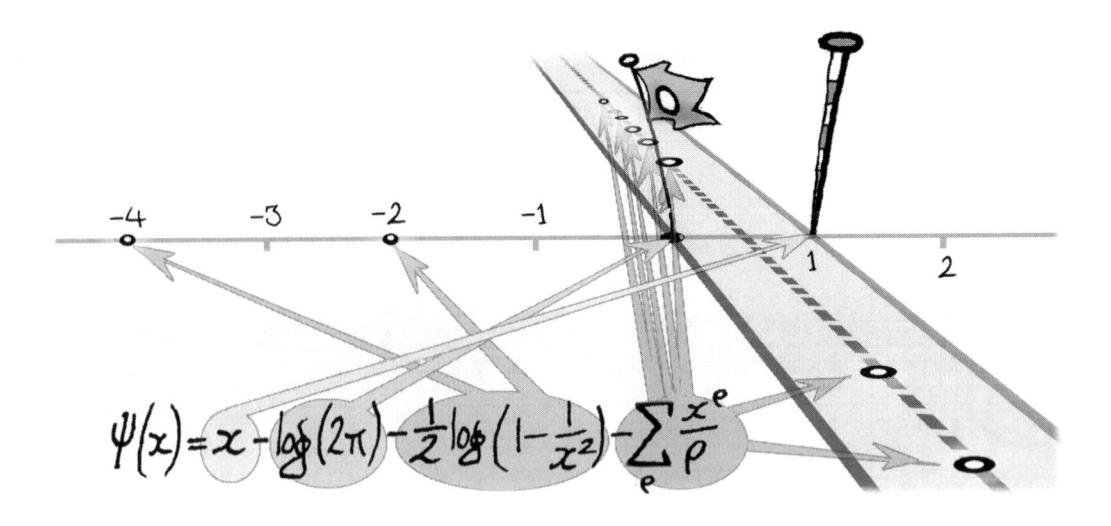

$$\psi(x) = x - \log(2\pi) - \tfrac{1}{2}\log\left(1 - \tfrac{1}{x^2}\right) - \sum_\rho \tfrac{x^\rho}{\rho}$$

The vertical axis here is not to scale: in reality, the nearest zeros to the horizontal axis are just over 14 units away, so the scaling here is roughly 10:1.

It's not quite as simple as each nontrivial zero producing a single spiral wave. Rather, it's *pairs* of nontrivial zeros which produce *pairs* of spiral waves. Remember that the zeros pair off in two ways: (vertically) in reflection across the real axis and (horizontally) across the vertical line through 1/2. If Riemann was correct in his belief that they all lie on the critical line, then the second of these pairings is irrelevant.

But we can't yet be sure that Riemann *was* correct, so it's still possible that this kind of situation might arise:

In such a situation, we'd have *two* pairs of nontrivial zeros. There would be a pair to the left of the critical line and a pair to the right of it (it's the vertical rather than the horizontal pairing which is relevant here). Each of these vertically reflected pairs of zeros would produce a pair of spiral waves.

The pairs of spiral waves produced by the first three pairs of zeros look like this:

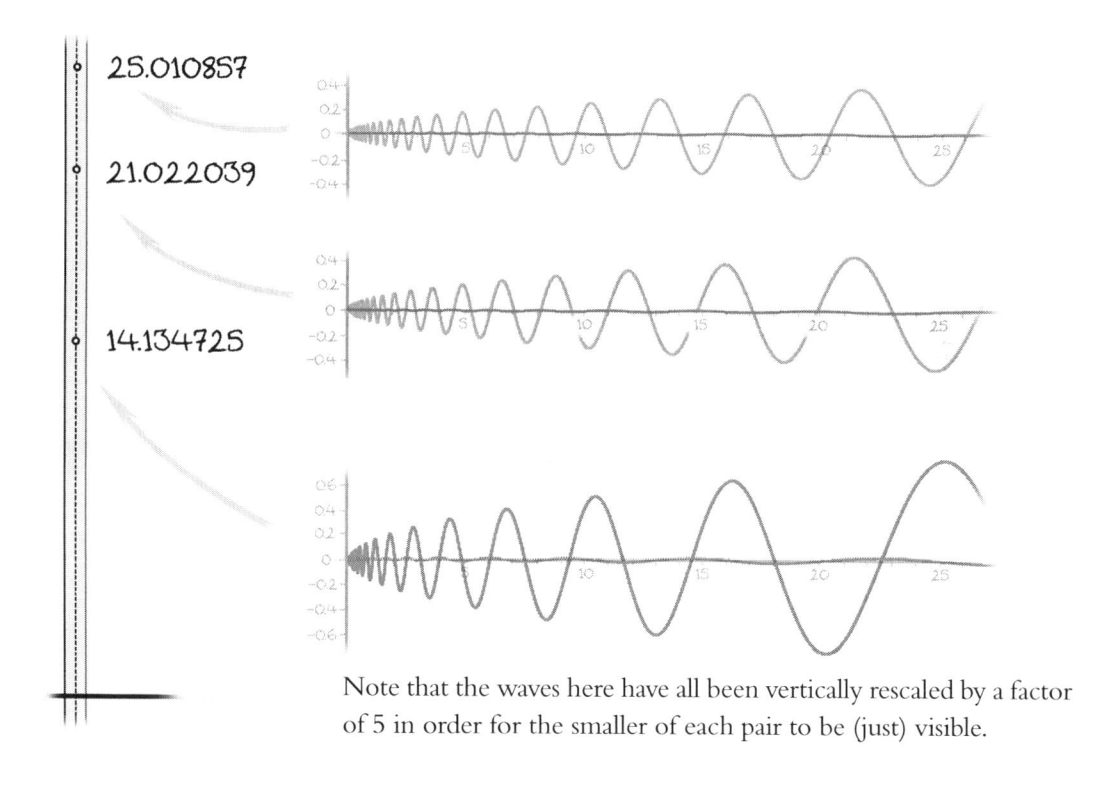

25.010857

21.022039

14.134725

Note that the waves here have all been vertically rescaled by a factor of 5 in order for the smaller of each pair to be (just) visible.

If you pay attention to where they cross the axis, you'll notice how the waves in each pair appear to have the same "frequency". Admittedly, we've not yet clearly defined this term for spiral waves, but the successive horizontal crossings of one wave in the pair correspond to the successive peaks of the other. Also (as all spiral waves do) both have growing amplitude, but one wave will always have a considerably larger amplitude if we compare corresponding peaks.

Before seeing exactly how the spiral waves relate to the pairs of zeros, we'll need to clear up this issue of the "specifications" of spiral waves and how they compare to the amplitude, frequency, *etc.* of ordinary sine waves, as outlined back in Chapter 13.

Unlike ordinary sine waves, spiral waves don't have simple wavelengths or frequencies. Remember, we can measure the wavelength of a sine wave by measuring the distance between its peaks, but a spiral wave's peaks become increasingly distant as we travel out along the number line. Similarly, we measure the frequency of a sine wave (which is directly related to its wavelength) in terms of how often, per unit of distance, the wave crosses the number line. But a spiral wave's crossings grow ever more distant, so there's no obvious, constant frequency. However, if we go back to the "Ferris wheel" visualisations of Chapters 13 and 14, whether we were looking at the circular or the spiral Ferris wheel, *the number of rotations per hour was fixed*. In the case of sine waves, this "speed-of-wheel-rotation" gives us the frequency of the wave, so we can similarly use the "speed-of-spiral-Ferris-wheel-rotation" to define the "frequency" of the spiral wave it produces (the repeated illustration on page 94 may help here).

Certainly, at a simple, visual and intuitive level, we can see that just as the sine wave on the left has a lower frequency than the sine wave on the right...

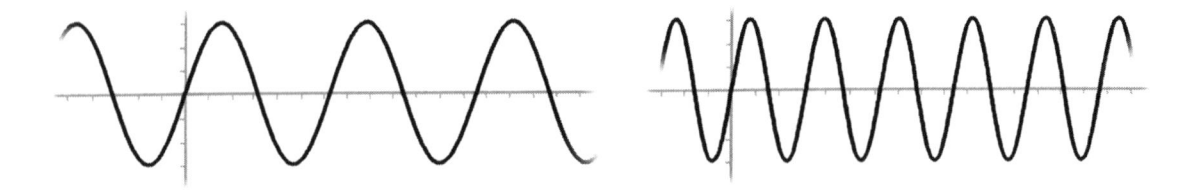

...the spiral wave on the left has a lower frequency than the spiral wave on the right (it's all about how often the horizontal axis is crossed):

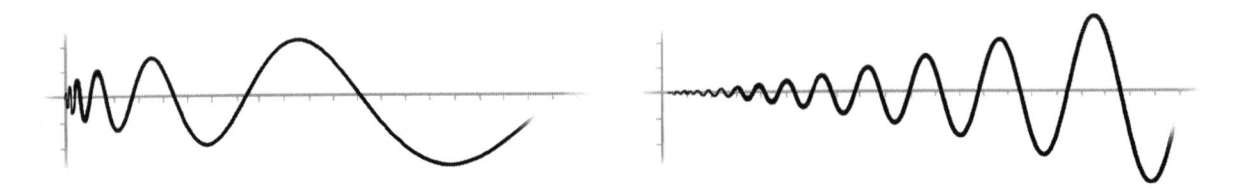

Next, we'll look at amplitude. An ordinary sine wave always has a fixed amplitude, which is just the height of each peak from the horizontal axis.

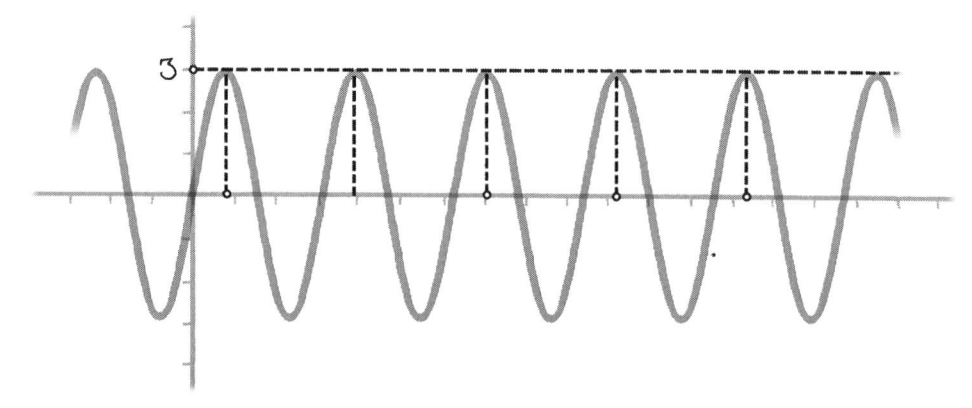

A spiral wave has no fixed amplitude, but does have an *amplitude growth rate*. All spiral waves have growing amplitude, but we can see that some grow faster than others. The (fixed) amplitude of a sine wave can be related back solely to the radius of the wheel in the (circular) Ferris wheel visualisation. The amplitude growth rate of a spiral wave, on the other hand, is affected by both the shape of the spiral involved in the (spiral) Ferris wheel visualisation and the rate at which the telescopic arm is cranked (the frequency). To find the amplitude growth rate, we multiply the frequency (rotations by hour) by the natural logarithm of the spiral's base.

One way to find the amplitude growth rate of a spiral wave (without access to the spiral that produced it) is to compare the positions and heights of successive peaks.

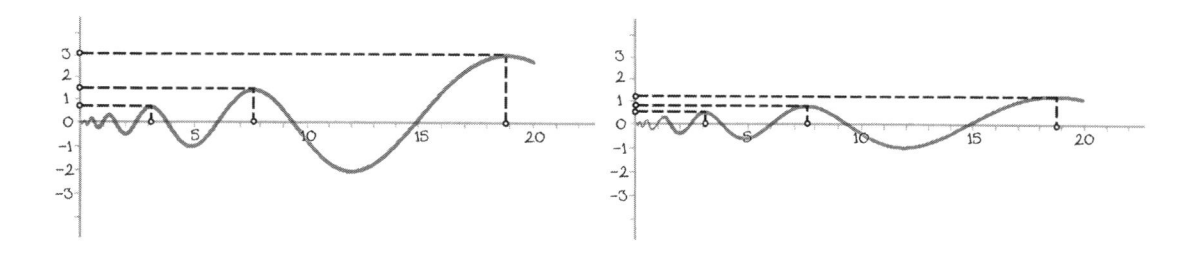

Both waves: the peaks marked are at approximately 3.071, 7.535 and 18.489. The corresponding peak heights are approximately 0.692, 1.419 and 2.909 (left) and 0.498, 0.780 and 1.222 (right).

89

It turns out that for any spiral wave, the position of each peak can be arrived at by multiplying the position of the previous peak by some fixed number. In the last illustration, it's 2.453... for both waves (we'll call that the "peak position factor"). Similarly, each peak height of a given spiral wave can be arrived at by multiplying the previous peak height by some fixed number (we'll call that the "peak height factor" of the wave): for the first of our waves, this is 2.050..., and for the second, it's 1.566.... To find the amplitude growth rate, we take the logarithm of the "peak height factor" and divide it by the logarithm of the "peak position factor".

For the first wave, that's the logarithm of 2.050... divided by the logarithm of 2.453..., which is 0.718... divided by 0.897..., which gives 0.8 – that's its rate of amplitude growth (all of these logarithms are "base-e" or "natural", by the way):

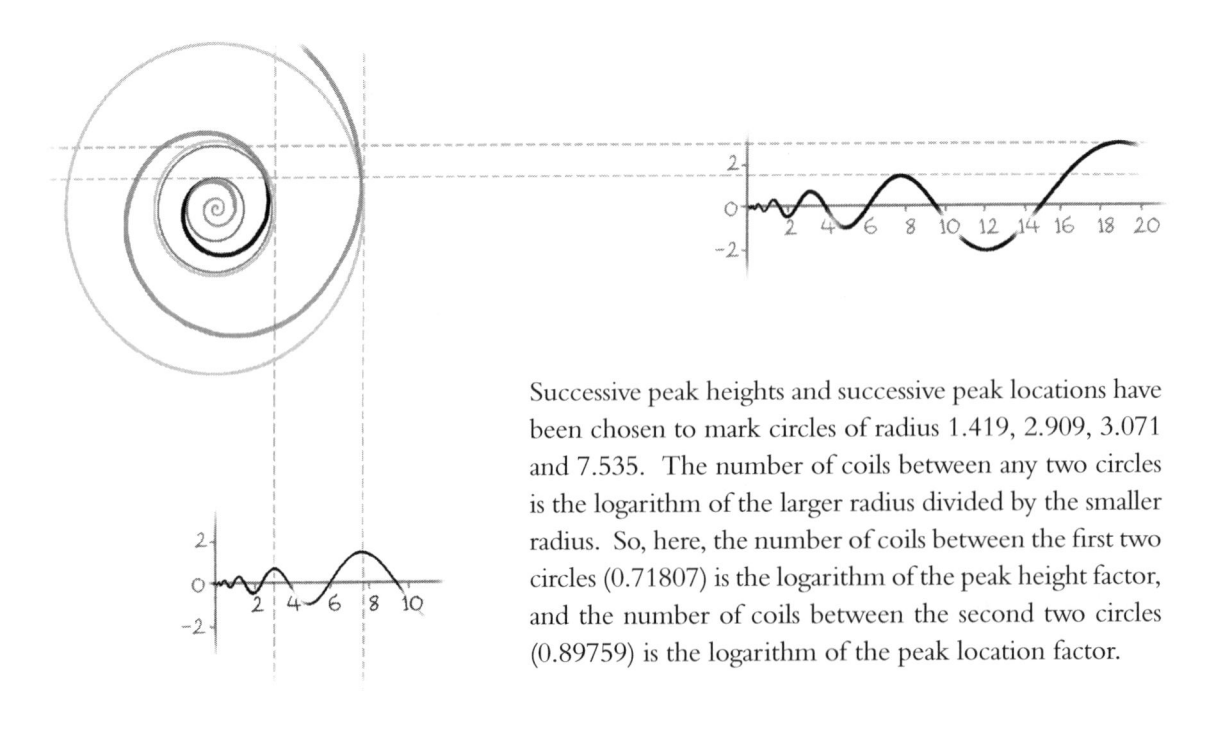

Successive peak heights and successive peak locations have been chosen to mark circles of radius 1.419, 2.909, 3.071 and 7.535. The number of coils between any two circles is the logarithm of the larger radius divided by the smaller radius. So, here, the number of coils between the first two circles (0.71807) is the logarithm of the peak height factor, and the number of coils between the second two circles (0.89759) is the logarithm of the peak location factor.

For the second wave, it's the logarithm of 1.566... divided by the logarithm of 2.453..., which is 0.448... divided by 0.897..., giving 0.5 – that's its rate of amplitude growth:

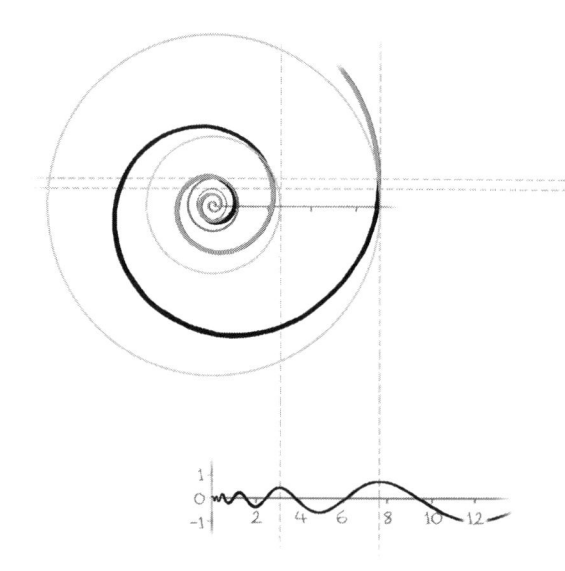

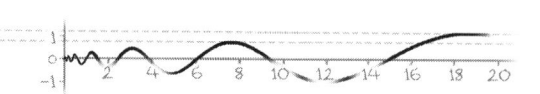

Here, successive peak heights are used to mark circles of radius 0.780, 1.222 (peak locations are still marking circles of radius 3.071 and 7.535). The number of coils between the first two circles (0.44879) is the logarithm of the peak height factor, and the logarithm of the peak location factor remains 0.89759.

It's possible for spiral waves produced by differently shaped spirals to have the same amplitude growth rates if their frequencies are chosen appropriately. But frequency and amplitude growth rate alone can't account for the fact that some spiral waves "maintain a bigger amplitude" relative to others. That is, two different waves can have the same amplitude growth rate and same frequency (this can only happen if they're produced by the same spiral) but one is always going to have a greater amplitude relative to the other.

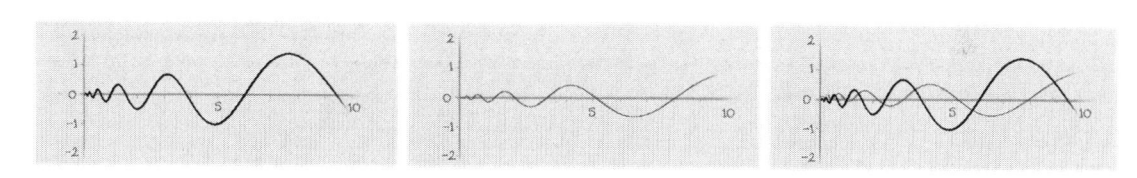

At certain points on the number line, the "bigger" wave will have amplitude 0 and the "smaller" waves will have a positive amplitude. But it's always possible to shift the waves so that the peaks and crossings line up, and then the "bigger" one *always* has greater amplitude.

You might remember that as well as amplitude and frequency/wavelength, ordinary sine waves have a third specification called *phase*. A sine wave shifted along the horizontal axis has a different phase (but the same amplitude and frequency/wavelength) as the original sine wave:

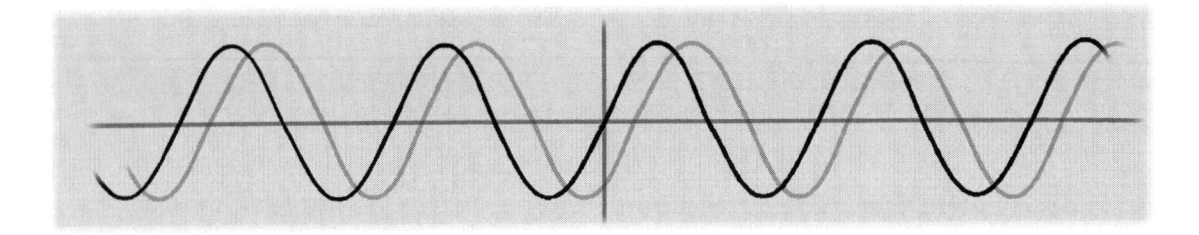

Spiral waves also have a third specification, closely related to phase, which I shall call the "phase point". Again, we refer back to our Ferris wheel visualisations. On page 256 of Volume 1, we saw how the phase of a sine wave can be related to the initial position of the carriage on the (circular) wheel. Although usually given as an angle (relative to the positive horizontal axis), the phase of a sine wave could also be given as a "phase point", a point on the wheel (the circle whose radius equals the amplitude of the wave). Generalising this to spiral waves turns out to be very helpful. If frequency and amplitude growth rate are fixed, then the shape of the spiral underlying a spiral wave is fixed. Given a "phase point" in the plane (anything other than 0), it's always possible to rotate the spiral (centred where the axes cross) to a position where it passes through this point. We now imagine starting the spiral Ferris wheel mechanism with the carriage starting on the spiral track where the phase point lies.

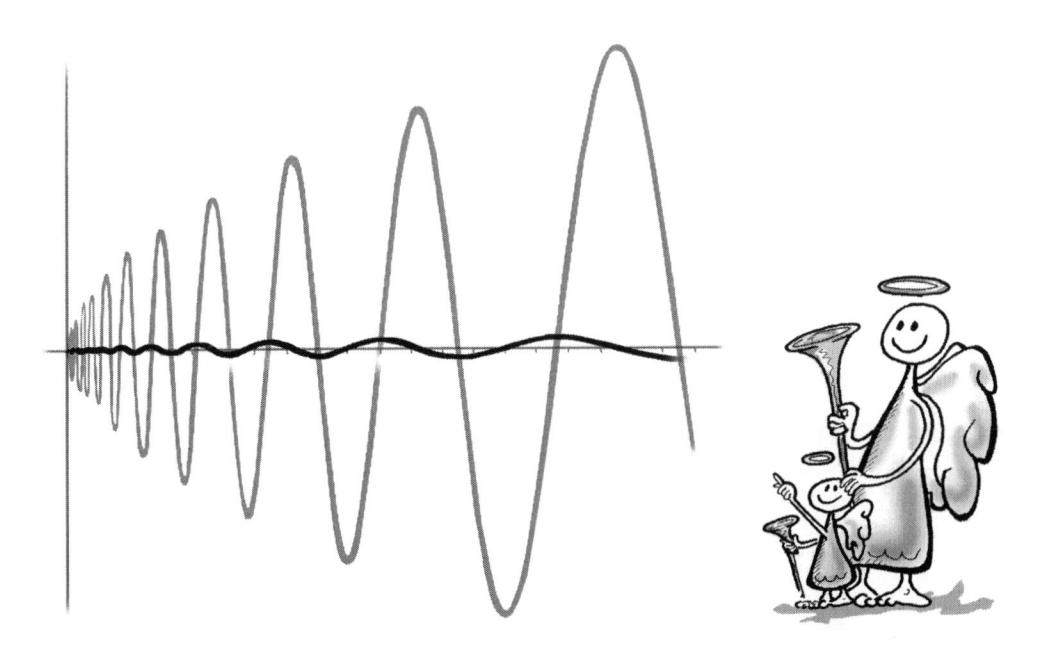

92

So we now have precise specifications to describe a particular spiral wave (frequency, amplitude growth rate and phase point). The next step is to describe a clear procedure which will take a pair of nontrivial zeta zeros (one above the real axis, one reflected below it) and use them to produce the relevant pair of spiral waves.

We'll make use of our spiral Ferris wheel visualisation, but first we need to know which spiral we'll be working with. This turns out to be easy, as we've met the procedure before. Draw a line in $\mathbb{C}$ through 0 and the upper nontrivial zero of the pair. This line defines an angle with the real axis which allows us to construct a unique equiangular spiral passing through 1.

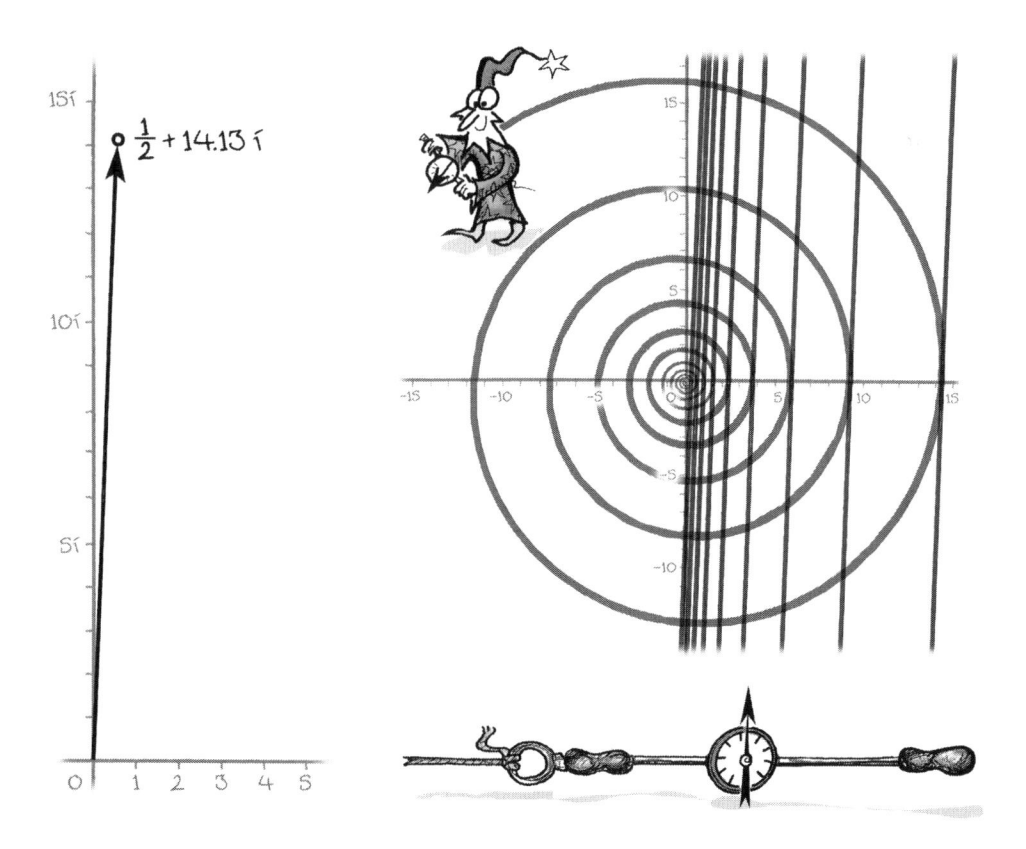

This spiral can then be used as the basis for a spiral Ferris wheel visualisation and the subsequent production of a spiral wave. Remember that the bicycle travelling along the number line always starts at 1 and accelerates in such a way that its speed

93

(in "units per hour") is equal to its position[2]. The frequency of the spiral wave that will result is just the number of rotations per hour on the spiral Ferris wheel. I've explained how the heights (imaginary parts) of the zeta zeros equal the frequencies of the spiral waves we need. In actual fact, heights must be divided by $2\times\pi$[3] (roughly 6.28). That's the number of rotations per hour we're going to set the wheel turning at and hence the frequency of the wave.

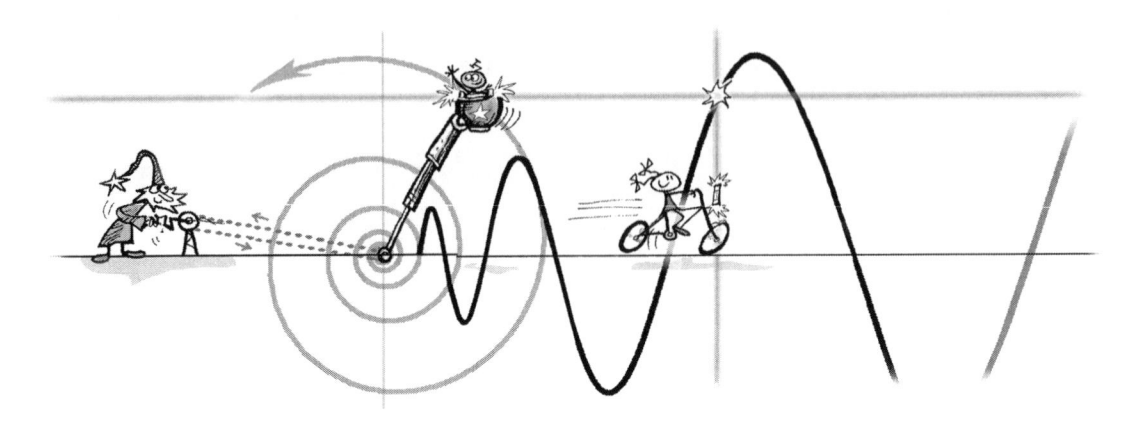

Remember that the carriage follows the spiral track, the telescopic arm being cranked at a constant number of rotations per hour (the "frequency" of the spiral wave). The bicycle accelerates so that its speed is always equal to its position. The points where the (horizontal) beam from the carriage crosses the (vertical) beam from the bicycle will trace out a spiral wave.

Notice how the starting position of the carriage on the spiral track is clearly going to affect the size of the amplitude at any given point on the resulting wave (imagine a range of starting positions, from close to the centre to huge distances out, and remember that the bicycle always starts at 1).

We're going to produce a pair of spiral waves – two waves with the same frequency and amplitude growth rate, but different phase points. To do this, we use the spiral and rate of rotation just described, but different starting positions for the carriage. One will be on the horizontal axis, which means the wave we'll get will pass through the horizontal axis at 1. The other will be on the vertical axis, which means the wave we'll get will have a peak at 1.

The first wave is produced as follows. Take two times the height of the upper zeta zero in the pair, divide that by its distance from 0 and divide the result by that same distance again. In the case of the zero 1/2 + 14.134725*i* (pictured on the right), we take 14.134725 and divide this by 14.1435657 (distance of the point from 0) to give 0.9993749, then again, to give approximately 0.07.

Our phase point is that far out along the positive horizontal axis. We rotate the spiral track until it passes through this point, attach the carriage there, and start the process. The apprentice on the bicycle rockets off along the positive horizontal axis, shining a vertical beam, and the carriage is set in motion at the appropriate number of rotations per hour (zeta zero height divided by $2\times\pi$), with the apprentice that's in it shining a horizontal beam.

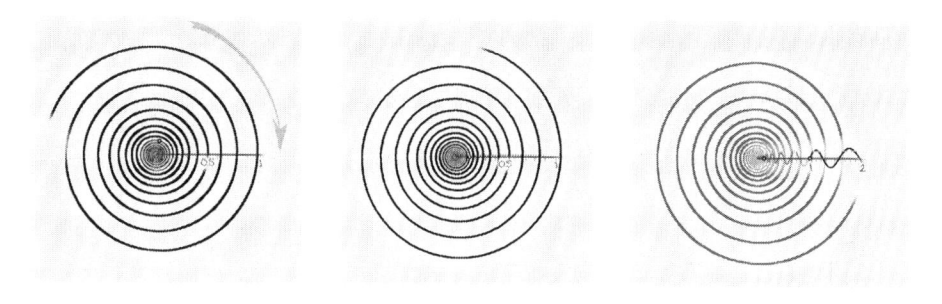

Locating the unique point and rotating the spiral as described, the resulting spiral wave will necessarily pass through the horizontal axis at 1.

We've already seen that this *kind of* process will produce a spiral wave. The important thing to grasp with this particular approach is that the exact specifications of the spiral wave we're going to get (its amplitude growth rate, frequency and phase point) are entirely determined by the pair of zeta zeros we start with.

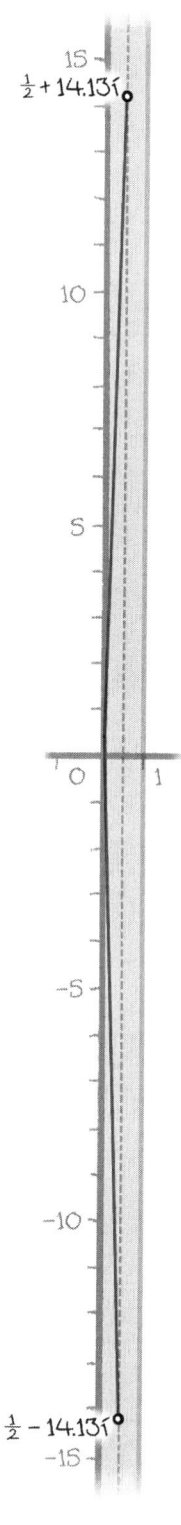

The second wave is produced in a very similar way to the first. The bicycle accelerates as usual and the speed of rotation of the carriage is the same as for the first wave, so the two waves have the same frequency (as in the illustration on page 87). The only difference is in the initial positioning of the carriage on the wheel. This time, we take two times the real part (horizontal position) of the nontrivial zeros (before we used two times the imaginary part, or height, of the upper zero), divide that by the distance of either of the zeros from 0 and then divide the answer by that same distance again. We again end up with a positive number, but a considerably smaller one than we got for the first wave. In the case of the zero $1/2 + 14.134725i$, we end up with 0.5 divided by 14.1435657 twice, which gives roughly 0.0025. Our point is then chosen that far up the positive vertical axis. The spiral is rotated until it passes through this point and then the carriage is attached there. The procedure then starts as before.

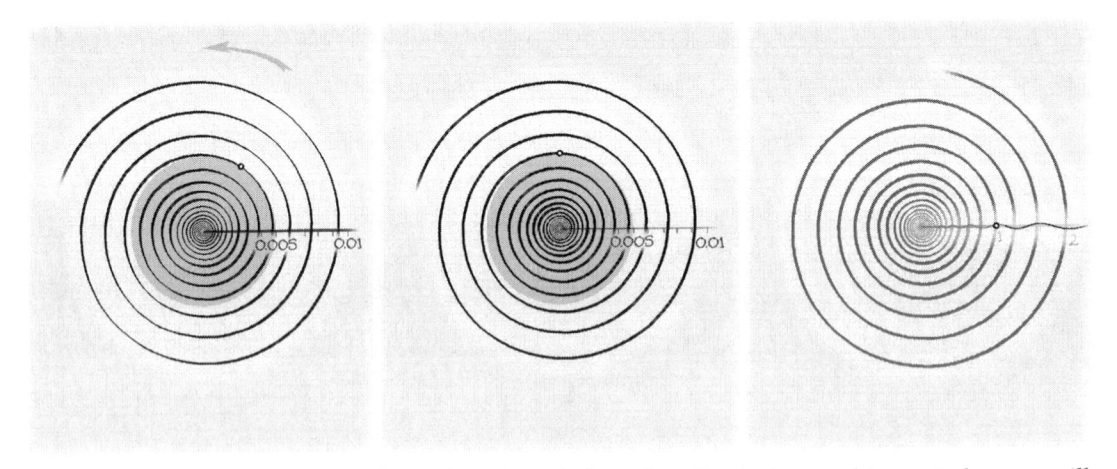

Locating the unique point and rotating the spiral as described, the resulting spiral wave will necessarily have a peak at 1 (barely visible here, due to the tiny wave amplitude).

This wave looks much smaller than the first one. Indeed, as we observed earlier, the waves in each pair share the same frequency, but one always has a much larger amplitude relative to the other. In fact, if we take the distance between the pair of nontrivial zeros in question, that's roughly how many times larger[4].

These two spiral waves, definitively produced by a pair of nontrivial zeta zeros, have not only the same frequency, but the same amplitude growth rate as well (recall that this is given by the frequency times the logarithm of the underlying spiral's base).

And helpfully, just as the imaginary part (height) of the upper zero in the pair gives us the frequency of the waves (once you've divided it by 2×π), the real part of either zero gives the amplitude growth rate.

The two waves have the same *rate* of growth, but the "bigger" one is, at any given point, in a more advanced stage of growth. To return to our musical analogy, this is like two notes getting louder at the same rate, but where one is already much louder than the other. In fact, if we look at the first pair of zeta zeros, which are at a distance of 14.134725... either side of the horizontal axis, the bigger of the associated waves is approximately 28.26955... (that's 2 × 14.1347235...) times "louder" than the smaller one at any given pair of corresponding peaks, but both have a growth rate of 0.5 because, like all known Riemann zeta zeros, this pair has horizontal position (real part) 0.5 in the complex number plane.

SUMMARY

Because the "primeness count staircase" can be exactly reconstructed in a way which, via Riemann's explicit formula, involves an infinite sequence of paired spiral waves, each wave can be thought of as contributing to the irregular distribution of prime numbers. The larger of the two waves in each pair can be thought of as "dominant" since its influence swamps the influence of the smaller wave. But both contributions are necessary for the whole scheme (as put forward by Riemann in his explicit formula) to work in the precise way that it does. Recall that:

✯ Each pair of waves is associated with a (vertically reflected) pair of nontrivial zeta zeros.

✯ The frequencies of both waves are controlled solely by the height (imaginary part) of the upper zeta zero of the pair.

✯ Their rates of amplitude growth are controlled solely by the real part of this zeta zero.

✪ The relative amplitudes of the waves are controlled by the relative distances of their phase points from 0. The positions of those phase points are determined by both the real and imaginary parts of the zeta zero.

chapter 21
the primes and the zeros

The role that the zeros of the Riemann zeta function play in Riemann's explicit formula (pages 82–86) should make it clear what was meant by the earlier claim that the zeta function's behaviour "encodes" the distribution of prime numbers.

So, the mystery of the distribution of primes has effectively been solved: it can be exactly reconstructed using the zeta zeros. But then there's the mystery of "where the zeta zeros come from". We know that they "come from" the zeta function's behaviour in $\mathbb{C}$, but we're still left with the lingering question "why *these particular* random looking numbers?" Also, as Enrico Bombieri has written:

> "*To me, that the distribution of prime numbers can be so accurately represented in a harmonic analysis is absolutely amazing and incredibly beautiful. It tells of an arcane music and a secret harmony composed by the prime numbers.*"[1]

That is, there's no obvious reason why the distribution of primes *should* give way to a harmonic decomposition like this – it comes as a total surprise! *And* it's widely perceived as something profoundly beautiful, once it has been grasped. As we'll see, this state of affairs concerning the number system and its hidden harmonic structure has inspired quite a bit of poetic, "ecstatic" and even religiously oriented language from serious minded mathematicians over the years.

To reiterate, we have found that lurking within (or behind) the number system is a very particular set of spiral waves (with seemingly random frequencies) which no one had the faintest idea were there until the end of the 1850s[2].

ANOTHER SUMMARY

People often express surprise when they learn that (via the spiral waves) there's an infinite sequence of these "zeta zeros" underlying the distribution of prime numbers. But, in one sense, it's not that surprising because there's a kind of circularity involved. We've seen how the sequence of primes can be used to define the Riemann zeta function, which leads us to the zeta zeros, which, in turn, can be used to reconstruct the sequence of primes.

To summarise:

✦ Starting with the primes, we can define Euler's zeta function.

✦ Riemann's zeta function then extends Euler's function from part of the number line to the complex number plane (apart from a single point).

✦ By considering the zeta function in this newly extended context, the nontrivial zeros become accessible.

Or, working in the other direction:

✦ We start with the nontrivial Riemann zeta zeros.

✯ For each pair of these (and we know there's an infinity of such pairs), we generate a pair of spiral waves, as described in the previous chapter.

✯ These, when appropriately combined, lead to a "staircase" graph, from which we can recover the entire set of prime numbers.

So it would be fair to say that the collection of primes and the collection of nontrivial zeros are "mutually generating" – they give rise to each other.

In fact, it's possible to produce a formula rather like Riemann's explicit formula, but "turned inside out", that is, a formula which describes a "nontrivial-zeros-counting function" (with its own staircase graph) in terms of the sequence of primes.

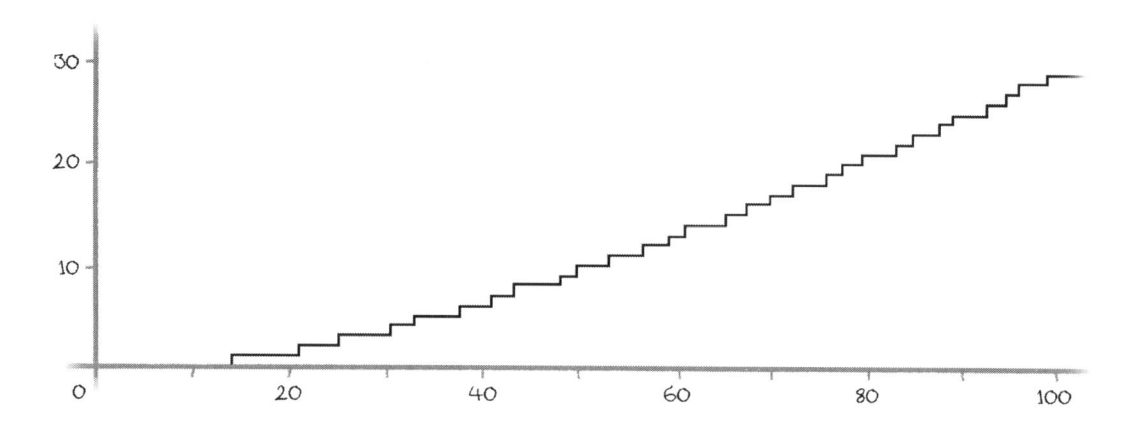

The vertical jumps in this graph occur at exactly those real numbers which are heights of Riemann zeta zeros (14.134..., 21.022..., 25.010..., 30.424..., 32.935..., etc.)

Rather like the nontrivial zeros contributing spiral waves in Riemann's formula, here, each prime number contributes a function, these all get added together, some extra bits get added on and we end up with a staircase function which can show us the exact locations of *all* nontrivial Riemann zeta zeros.

The formula looks like this[3] (but we won't worry about the details):

$$\mathcal{N}(T) = \frac{T}{2\pi}\log\frac{T}{2\pi} - \frac{T}{2\pi} - \frac{1}{\pi}\lim_{N\to\infty}\left[\sum_{p\leq N}\sum_{k=1}^{\lfloor\frac{\log N}{\log P}\rfloor}\frac{\sin(kT\log P)}{kp^{k/2}} - \right.$$

$$\int_1^N \frac{\sin(T\log t)}{\sqrt{t}\log t}\,dt - \frac{\sin(T\log N)}{\log N}\left\{\sum_{p\leq N}\sum_{k=1}^{\lfloor\frac{\log N}{\log P}\rfloor}\log p\cdot p^{-k/2} - 2\sqrt{N}\right\}\Big]$$

$$+ \frac{1}{2\pi}\left(\arg\Gamma(\tfrac{1}{2}+iT) - T\log T + T\right) + \frac{1}{\pi}\tan^{-1}(2T) - \frac{1}{4\pi}\tan^{-1}(\sinh\pi T)$$

All that matters is that you can both (1) use Riemann's explicit formula and knowledge of zeta zeros to reconstruct the sequence of primes or (2) use this other formula and knowledge of the primes to reproduce the sequence of nontrivial zeta zeros.

In this light, the prime numbers and the nontrivial zeros of the zeta function can be understood as being like two sides of a coin. The big question we'll eventually be exploring is *what exactly is that coin?*

This relationship between the prime numbers and the nontrivial zeros, which we see embodied in Riemann's explicit formula and in the formula illustrated above, is understood by mathematicians as an example of what's called a *Fourier duality*. The primes and zeros are said to be *dual* in this sense. You might remember a mention back in Chapter 13 of "Fourier analysis", the branch of mathematics (often applied to physics) which concerns the breaking down of compound waveforms into the sine waves from which they're built. Fourier duality comes into this and various instances of it can be found at all levels of physical reality[4]. The thing to bear in mind is that *this* instance of Fourier duality emerges directly from the number system, while all others require more advanced mathematical models (unavoidably founded on the number system) in order to be identified or described. So, in some sense, the Fourier duality which exists between the zeros and primes could be thought of as the "oldest", most "primal" or "archetypal" of all Fourier dualities.

Despite this "duality" or "mutually generating relationship" between primes and zeros, it would be tempting to argue that the primes are *more fundamental*. They're nice, simple counting numbers, whereas the zeros lie in awkward positions in the complex number plane involving quantities like 14.134725..., 21.022040... and 25.010858... You can easily explain prime numbers to children, but not (at least within our current framework of understanding) the nontrivial zeros. The primes have been known about for thousands of years and can be related to the everyday experience of "dividing things into equal amounts". The nontrivial zeros have been known about since only the late 1850s and cannot in any way be related to our daily lives (unless we happen to be analytic number theorists).

Still, I would argue that the zeta zeros are at least as fundamental as the primes and, in fact, that it's probably most helpful to think of the zeros as "underlying" the distribution of primes, lying hidden beneath it, waiting to be discovered, until Riemann came along. They are undeniably a key part of the deep infrastructure of the number system and therefore of our perception and understanding of reality, again bearing in mind this quotation:

> *"As archetypes of our representation of the world, numbers form, in the strongest sense, part of ourselves, to such an extent that it can legitimately be asked whether the subject of study of arithmetic is not the human mind itself. From this a strange fascination arises: how can it be that these numbers, which lie so deeply within ourselves, also give rise to such formidable enigmas? Among all these mysteries, that of the prime numbers is undoubtedly the most ancient and most resistant."*
>
> G. Tenenbaum and M. Mendès France [5]

It's worth pointing out that the locations of the nontrivial zeros are, like the locations of the primes, not the result of anyone's theories or opinions, they're *just like that*. In a sense, when we look at the nontrivial zeros, we're just looking at the distribution of primes from a different perspective.

As we've seen, the primes are at the core of the system of counting numbers. Behind their seemingly jumbled arrangement is a harmonic structure – what has been described as a kind of music. The nontrivial zeros are something like the notes which make up this music.

Riemann, recall, produced an *exact formula* directly relating the primes and the nontrivial zeros, with the primes on one side of the equation and the zeros on the other. His explicit formula is one of the greatest achievements of human rational thought, for it demonstrated conclusively that the number system has a (previously unimagined) "harmonic infrastructure".

Because Riemann was able to produce an exact formula, any talk of the "random" or "unpredictable" nature of the primes only goes so far. There's a precise, deterministic relationship between the primes and the zeta zeros. But there's also a "knowledge problem": as finite, mortal beings, we can never "know" all of the zeta zeros at once.

So, the question of whether or not there's a "pattern" in the primes is seen to be much less simple than we might initially have supposed. The existence of the nontrivial zeros and their associated spiral waves underlying the sequence of primes is arguably a pattern of sorts, but not in a sense that involves any kind of repetition or predictability.

We'll now quickly review what's known about the zeros of Riemann's zeta function:

✫ There are zeros at all of the negative even integers $-2, -4, -6, -8, -10, \ldots$ (called *trivial zeros*). All others are known as *nontrivial zeros*.

✫ All nontrivial zeros exist in the vertical strip of $\mathbb{C}$ enclosed by the vertical lines through 0 and 1 (called the *critical strip*).

✯ There are infinitely many nontrivial zeros.

✯ Every nontrivial zero above the horizontal axis has a "partner" below the horizontal axis, its "vertical mirror image", directly below it and the same distance from the horizontal axis.

✯ The first few (working our way up from the horizontal axis) are found on the *critical line* which passes vertically through 1/2, at the following distances above the horizontal axis: 14.134725..., 21.022039..., 25.010857..., 30.424876..., 32.935061..., 37.586178...

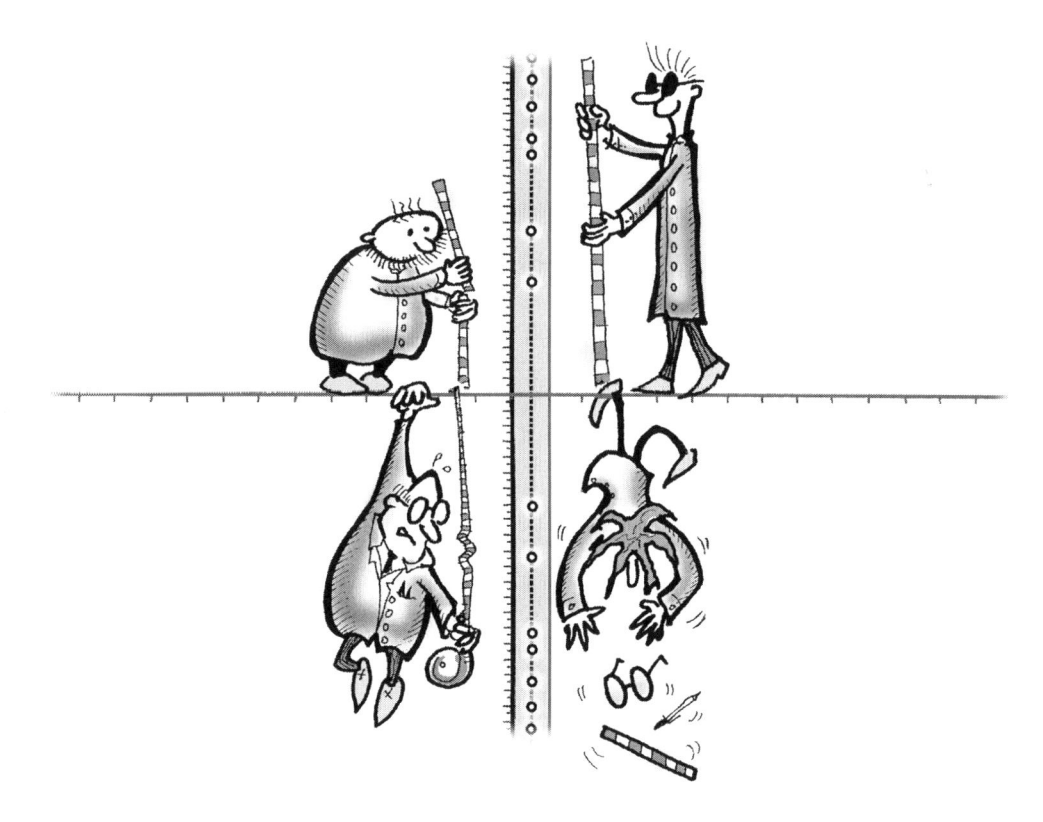

✯ Every zero also has a reflection across 1/2 and, as a result, a "horizontal mirror image" partner, reflected across the critical line. If a zero lies *on* the critical line, then it's its "own horizontal reflection", so there's no real need for a partner of this type. The next illustration should make this clear:

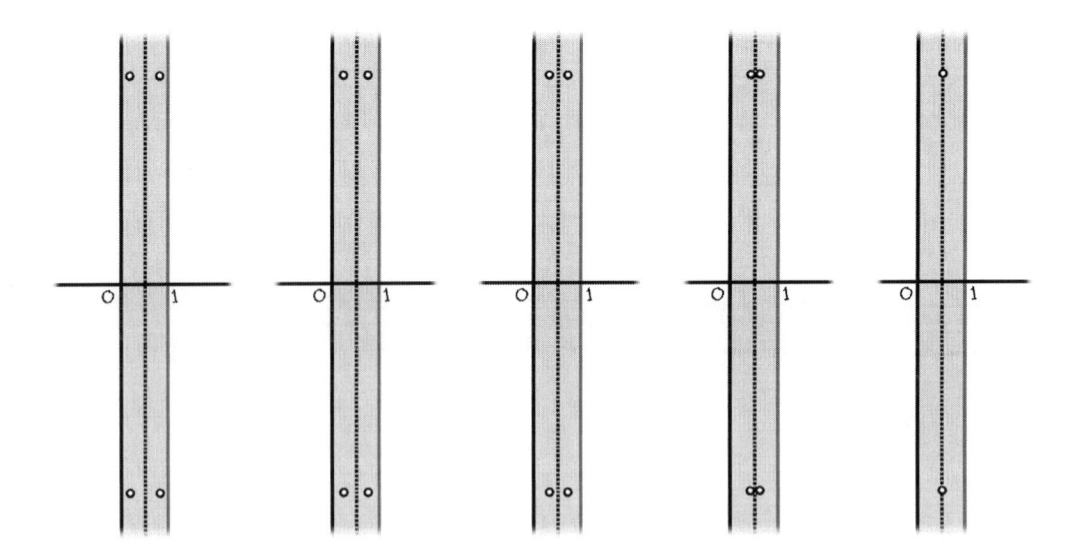

✭ All known zeros of the zeta function lie on the critical line (paired off above and below the horizontal axis). As of the early 21st century, there are billions of pairs known.

✭ Rather like the way the primes appear as a random looking jumble on the number line but gradually tend to thin out "on average" according to a formula involving logarithms, the nontrivial zeros appear as a jumble along the critical line, but rather than thinning out, they tend to squash together more and more the farther up the line you look, also according to a logarithmic formula.

✭ If one nontrivial zero were found to lie *off* the critical line, we would immediately have *four* zeros off the line (see page 77), instantly disproving Riemann's conjecture that all nontrivial zeros lie on the critical line.

✭ Infinitely many nontrivial zeros lie on the critical line (this was proved by Godfrey Hardy and John Littlewood in 1921[6]). Note that this *doesn't* rule out infinitely many nontrivial zeros also lying *off* the critical line.

✭ We can also be sure that at least 40% of the nontrivial zeros lie on the critical line (this was proved by J. Brian Conrey in 1989[7]).

THE CALCULATION OF RIEMANN ZEROS

Riemann calculated the first few pairs of nontrivial zeros of his zeta function and, finding them all to lie on the critical line, hypothesised that *all nontrivial zeros* would lie on this line. It's not entirely clear why he felt so confident about this, although certain lines of reasoning have been suggested.

It occurs to some people to ask how, if Riemann was working within a limited range of computational accuracy, he could be so sure that any of his zeros lay *exactly* on the line, that is, that they had real part exactly equal to 1/2. Thinking about it, even if you were to compute the horizontal location (real part) of a zero to ten decimal places and find it to be 0.5000000000, this is not enough, for the actual position might be 0.5000000000000001 ("close to" the critical line by most people's standards, but most definitely not *on* it). To answer this question, we must recall that a zero off the line will be horizontally mirrored by another zero across the critical line. In other words, if there were a zero with horizontal position 0.50000000000000001, then there would also necessarily be a zero at the same height with horizontal position 0.49999999999999999. Riemann used a technique which was able to detect the *number* of zeros in small regions of the complex plane and, in this way, by successfully detecting one zero rather than two in a carefully chosen and appropriately tiny region of the plane, he was able to show that his computed zeros were *exactly* on the line.

The next serious data on the nontrivial zeros appeared in 1903 when Jørgen Gram published a paper containing the heights of the first fifteen zeros, with the first ten given to six decimal places [8]. This was improved in 1914 by Ralf Backlund, who successfully calculated all nontrivial zeros with heights less than 200 [9]. In 1925, John Hutchinson took this up to 300 [10]. These researchers used a method called the *Euler–Maclaurin summation formula* which had emerged in the early 18th century, and which Euler himself had used to calculate values of his zeta function at all positive integers from 2 to 16.

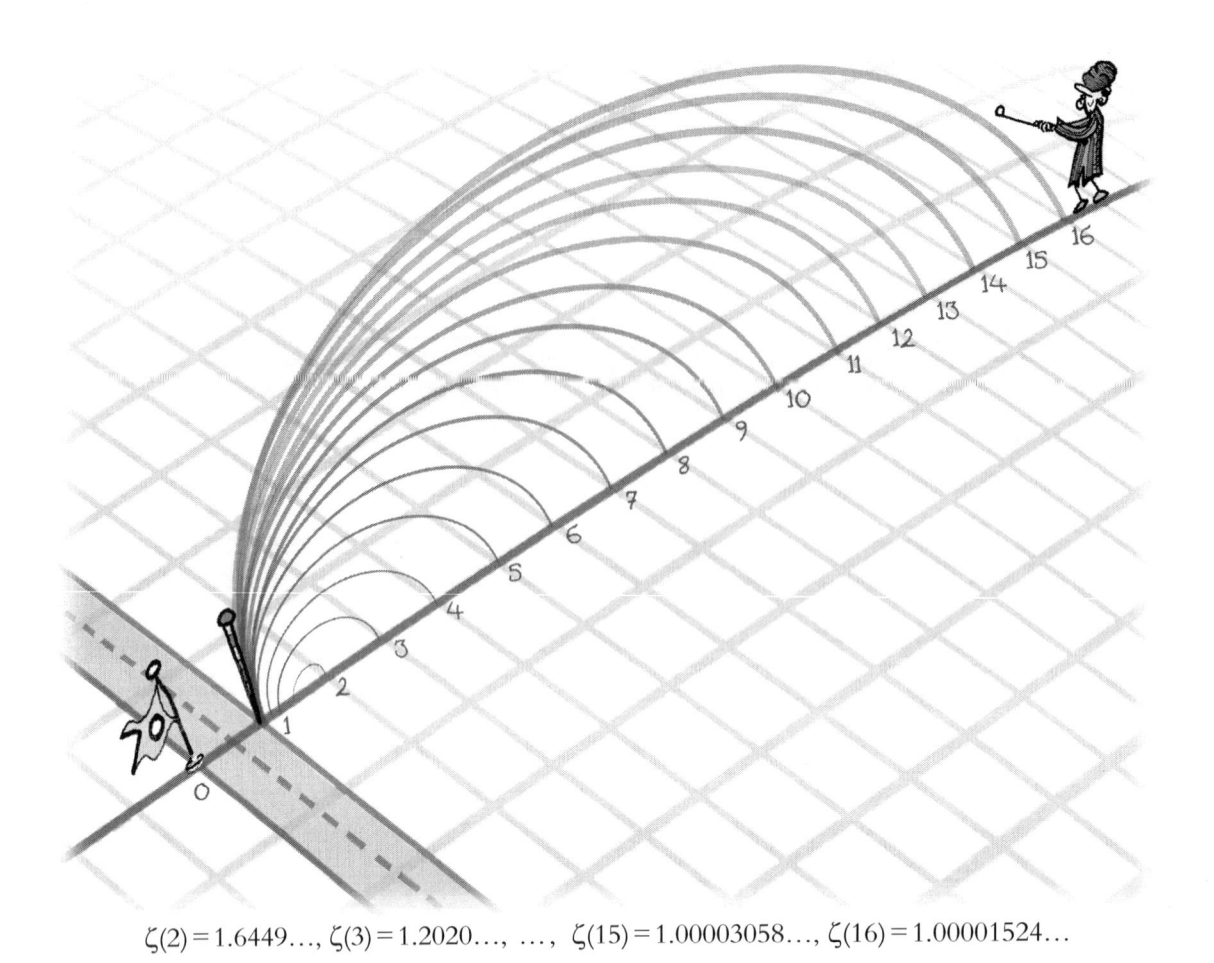

$\zeta(2) = 1.6449\ldots,\ \zeta(3) = 1.2020\ldots,\ \ldots,\ \zeta(15) = 1.00003058\ldots,\ \zeta(16) = 1.00001524\ldots$

Not surprisingly, the calculation of zeta zeros is tedious, time-consuming and not hugely inspiring. In other words, it's just the sort of thing powerful computers are good for. Alan Turing, pioneer of computational theory and code-breaking WWII hero, anticipated this in 1939, at a time when sufficiently powerful computers were not yet available:. Turing, then based at Trinity College, Cambridge, became interested in the zeta function and designed a mechanical device involving many toothed gear wheels for computing its zeros [11]. Cambridge University technicians set to work constructing this for him, but the machine was never completed. Turing's main motivation seems to have been to find a nontrivial zero off the critical line and therefore disprove Riemann's conjecture that they all lie *on* it. Now, with the benefit of early 21st century electronic computer technology, we know the locations of billions of zeta zeros up the critical line and it's clear that Turing's mechanical approach was doomed to fail – the limits of precision which restrict any such mechanical device (however precisely engineered) would never have allowed computations anywhere near such heights on the critical line.

I should stress that no amount of computation can ever *prove* Riemann's conjecture. However many pairs of zeros you confirm lie on the critical line, you can never be certain that the next pair will. Despite the impressive numerical evidence, a disproof of Riemann's conjecture is still theoretically possible – although the great majority of mathematicians believe that this won't happen, that Riemann was correct.

Here's a compact historical overview of the work that's been done on computing pairs of nontrivial zeros:

year	number of pairs	person(s) responsible for calculations
1903	15	J.P. Gram
1914	79	R.J. Backlund
1925	138	J.I. Hutchinson
1935	1041	E.C. Titchmarsh
1953	1104	A.M. Turing
1956	15 000	D.H. Lehmer
1956	25 000	D.H. Lehmer
1958	35 337	N.A. Meller
1966	250 000	R.S. Lehman
1968	3 500 000	J.B. Rosser, J.M. Yohe and L. Schoenfeld
1977	40 000 000	R.P. Brent
1979	81 000 001	R.P. Brent
1982	200 000 001	R.P. Brent, J. van de Lune, H.J.J te Riele and D.T. Winter
1983	300 000 001	J. van de Lune and H.J.J. te Riele
1986	1 500 000 001	J. van de Lune, H.J.J. te Riele and D.T. Winter
2001	10 000 000 000	J. van de Lune
2004	900 000 000 000	S. Wedeniwski
2004	10 000 000 000 000	X. Gourdon and P. Demichel

Conspicuously absent from this list is the name "A. Odlyzko". Andrew Odlyzko (currently at the University of Minnesota) is the name now most widely associated with the computation of Riemann zeta zeros. In 1978, when he had access to one of the first Cray supercomputers, he began calculating some of the previously computed zeros to much higher precision. Over the years, he's gone on to compute billions of zeros, but rather than focussing on the initial stretch, he's been more interested to see what's going on much farther up the critical line. Titles of two of his papers reporting his findings are "The 10^{20}-th zero of the Riemann zeta function and 175 million of its neighbors"[12] and "The 10^{22}-nd zero of the Riemann zeta function"[13] (the latter considers *10 billion* neighbours – and in case you didn't know, "10^{22}" means the number "1 followed by twenty-two 0's", that is, 10000000000000000000000). Odlyzko has also contributed significantly to the field by developing new, improved algorithms (computational procedures). His contributions to this area of research go much further than simply confirming that large numbers of nontrivial zeros are on the critical line, though, as we we'll go on to see in Volume 3.

In 2002, an Internet-based "distributive computation" project called "ZetaGrid" was launched by Sebastian Wedeniwski[14]. This was along similar lines to the "Great Internet Mersenne Prime Search" (GIMPS), a project for finding new prime numbers which has been running since 1996. Hundreds of thousands of volunteers worldwide subscribed to a scheme allowing their personal computer's processor to be used as part of a mass computation whenever their machine was connected to the Internet and otherwise idle. By the time the project finished in 2005, the locations of 100 billion pairs of zeros had been computed. I don't think anyone was that surprised that they were all confirmed to lie on the critical line.

THE ZEROS AND THE "DEVIATION"

Because the horizontal position (or real part) of a pair of nontrivial zeros is the same as the amplitude growth rate of the associated pair of spiral waves, the possibility that there might be nontrivial zeros off the critical line has major implications. For if there's a pair of zeros with real part even the tiniest amount greater than ½...

0.51?

0.501?

0.500001?

0.50000000000000000000000000001?

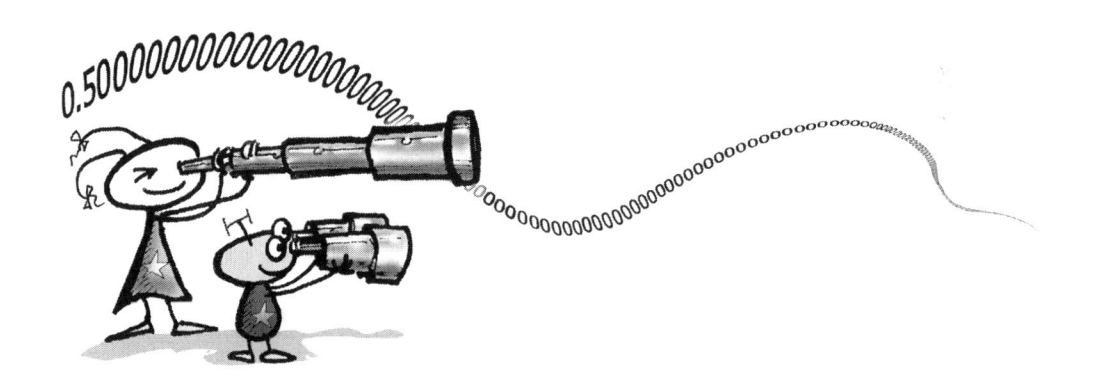

0.500000000000000000000000000000000000000000000001?

0.5000000000000000000000000000000000000000000...?

...then we'll end up with a spiral wave which starts off behaving similarly to all the others – you wouldn't notice anything unusual about it in the initial stretch of number line from 0. But, the farther you look down the number line, the more the difference in amplitude will get magnified, so if you're prepared to look far enough down the line, this wave's peaks will become a million, trillion (or whatever) times higher than those of the others produced by the "real part ½" Riemann zeros (the ones on the critical line). If, on the other hand, all the zeros are on the critical line as Riemann speculated, then although the peak heights of the various spiral waves involved in the decomposition will vary (relative to each other) in any given chunk of the number line, all of their *relative* magnitudes will remain the same, because the waves' *amplitude growth rates* will be the same.

So, all of the nontrivial zeros being on the critical line would mean one thing, and just one pair (or pair of pairs, as it would have to be) slightly off the line would mean something very different.

Enrico Bombieri has written that if there *is* a pair of nontrivial zeros off to the right of the critical line[15], then:

> "*In an orchestra, that would be like one loud instrument that drowns out the others – an aesthetically distasteful situation.*"[16]

These spiral waves are all getting added together to create something that describes exactly where the prime numbers are. So, each one makes a contribution, like this:

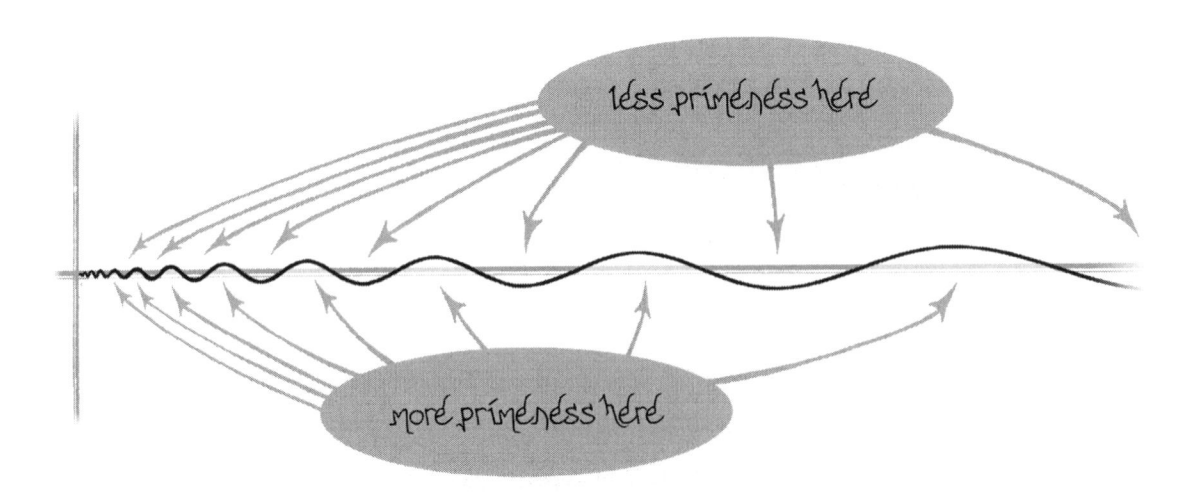

less primeness here

more primeness here

This means that the spiral waves produced by any pairs of zeros off to the right of the critical line would "slosh the primeness around" much more vigorously than all of the others, ultimately achieving a kind of dominance over their collective effect. To put it crudely, this would "mess up the regularity of the primeness distribution".

As Peter Sarnak puts it, if this is the case...

"...then the world is a very different place. The whole structure of integers and prime numbers would be very different to what we could imagine. In a way, it would be more interesting if it were false, but it would be a disaster [for mathematics.]*"*[17]

Bombieri adds:

"Even a single exception to Riemann's conjecture would have enormously strange consequences for the distribution of prime numbers."[18]

We've considered the deviation between the actual amount of "primeness" and the amount predicted by a simple formula. This deviation is describing the way in which (loosely speaking) the primes fluctuate around an average behaviour. In some stretches of the number line there are more prime numbers than predicted, in other stretches there are less. This is what Riemann's research was effectively probing. And, remarkably, he was able to show that these fluctuations are built out of an infinite stack of spiral waves.

This deviation could also be thought of as the "error" in the Prime Number Theorem. The PNT, remember, tells us that the amount of primeness (see Chapter 12) between 0 and a point on the number line is approximately equal to the location of that point (the number it represents). So, the amount of primeness less than 2197 is approximately 2197 (in actuality, it's 2182.273...). The prediction will never be perfectly accurate, so there's always an error (or deviation). That's what this graph is showing, recall:

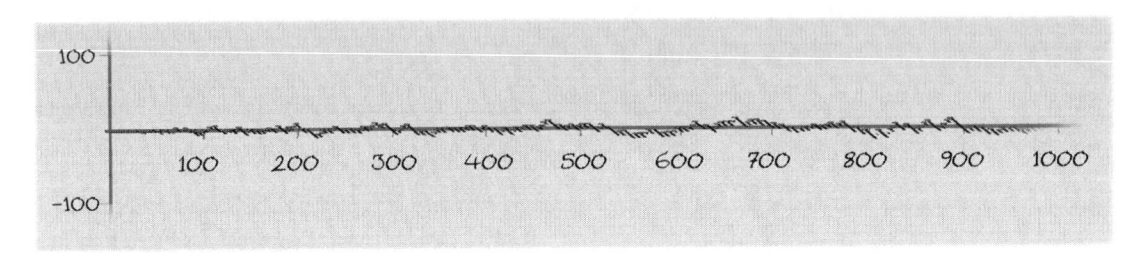

The deviation graph looks pretty unimpressive on this range, which is why it's usually presented in vertically rescaled form, as in the picture below, opposite.

The farther off to the right of the critical line a pair of nontrivial zeros were to stray, the bigger the amplitude growth rate of the corresponding pair of spiral waves would be, and so the bigger their contribution to the PNT error. In this way, the size of the PNT error can be directly related to how far any nontrivial zeros stray from the critical line.

If Riemann was correct that the zeros are all on the line, then all of the spiral waves involved will have rate of amplitude growth ½, none will become dominant, and the PNT error will be as small as possible. That's one extreme scenario. The opposite extreme would involve nontrivial zeros as far from the critical line as possible, that is, right out on the edge of the critical strip (on the lines through 0 and 1).

So the size of the PNT error can be linked to something called the *zero-free region*. This is just a pair of strips within the critical strip like this...

...where the inside edges of the strips are determined by the farthest horizontal-straying nontrivial zeros. The first extreme scenarios would be that the zero-free region is the whole critical strip except the critical line, and its opposite would be that *there was no* zero-free region (because of zeros lying out on the edge of the critical strip).

You may recall that the PNT had been proposed years before Riemann's birth and was still unproved at his death. I've stated it in quite loose terms, involving the word "approximately" ("the amount of primeness less than a given number is approximately that number"). And "approximately" is, of course, very much open to interpretation. But mathematicians have a number of very precise formulations which can be used to state conjectures and theorems in terms of approximation. Once again, we needn't worry about the details, but if you want to know, have a look in Appendix 14 (otherwise, you'll just have to accept that the PNT is stated in these precise terms). Now, however you interpret "approximately", you ought to agree that it's got something to do with the size of the error being not too big. Certainly the size of the error should be involved in some way. So the PNT should be somehow related to the size of the deviation:

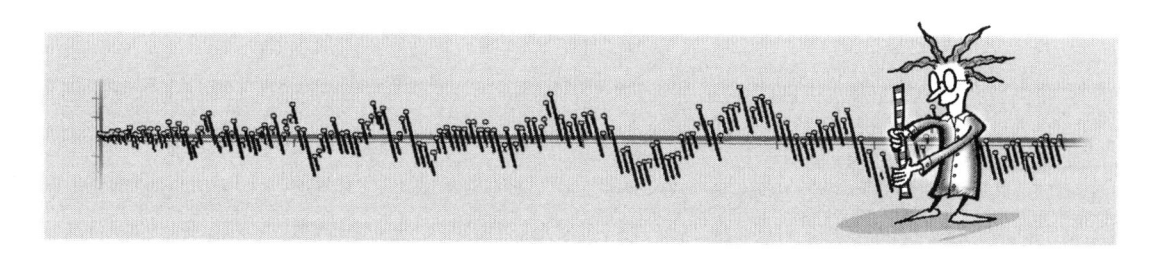

This deviation graph, we've seen, is built out of spiral waves linked to nontrivial zeta zeros. Once you've come to understand Riemann's explicit formula, it becomes quite easy to see that the truth of the PNT is mathematically equivalent to the following statement:

No nontrivial zeros of the zeta function lie on the vertical lines through 0 and 1.

That is, none of them lie on the edge of the critical strip – they must all be contained within its interior. This keeps the spiral waves involved from growing too fast, which keeps the overall deviation from growing too fast. That, in turn, means that the PNT error must be "bounded" in such a way that the PNT is true. In 1896, thirty years after Riemann had died, Jacques Hadamard and Charles de la Vallée Poussin separately proved the truth of the above statement using some intricate analytic number theory and, in the process, proved the PNT.

THE ZEROS AND THE "DEVIATION"

We've learned about the existence of the complex number plane $\mathbb{C}$. We've also learned that mathematicians study functions on both the real number line $\mathbb{R}$ and on the complex plane. There's one function on $\mathbb{C}$ which is of particular importance in the study of prime numbers – the *Riemann zeta function*. We learned how this came out of Riemann extending Euler's earlier zeta function which was defined on half of $\mathbb{R}$ and was the subject of his "product formula". Euler's product formula, which relates the sequence of primes and the system of counting numbers, can be understood as a kind of extension or reinterpretation of the Fundamental Theorem of Arithmetic (see page 46).

My feeling (and this is no longer actual mathematics, but rather my feelings about a certain area of mathematics) is that there is a particularly deep, subtle phenomenon associated with what we know as the sequence of prime numbers: Euclid looked at this through the "prism" of the counting numbers and saw the FTA (around 300BCE),

Leonhard Euler looked at the same thing through the prism of the number line and saw his product formula (around 1737), then Bernhard Riemann looked at it through the prism of the complex plane and saw his zeta function (1859).

In each case, a new layer of understanding was opened up in connection with prime numbers. Initially, they were just experienced as individual anomalies, counting numbers which were distinguished by their "indivisibility". Euclid, working with the counting numbers in ancient Greek times, noticed something which we now know as the FTA and thereby extended humanity's understanding of the primes – they were revealed as "building blocks" of the counting numbers. Euler then extended our understanding, effectively re-expressing the FTA in the context of his product formula, which "lives" on that part of the number line to the right of 1 (the counting numbers are seen as isolated points in this continuous half-line).

Riemann further extended the theory of prime numbers by extending Euler's product formula to the complex plane $\mathbb{C}$, of which the number line is just one tiny "slice". In this way, Riemann was able to reveal the spiral waves we've been looking at. This curious collection of "tones", which underlies the irregular spacings and seeming lack of pattern associated with the prime numbers, was entirely unknown to the mathematical community prior to 1859.

It's tempting to imagine that a further extension of our knowledge could reveal the same truth at yet another, deeper level, through a new and as yet unimagined numerical "prism". I may be wrong, but certain occurrences in the mathematical sciences do seem to be suggesting that something like this is beginning to emerge. There are some very interesting signs, as we'll see in Volume 3.

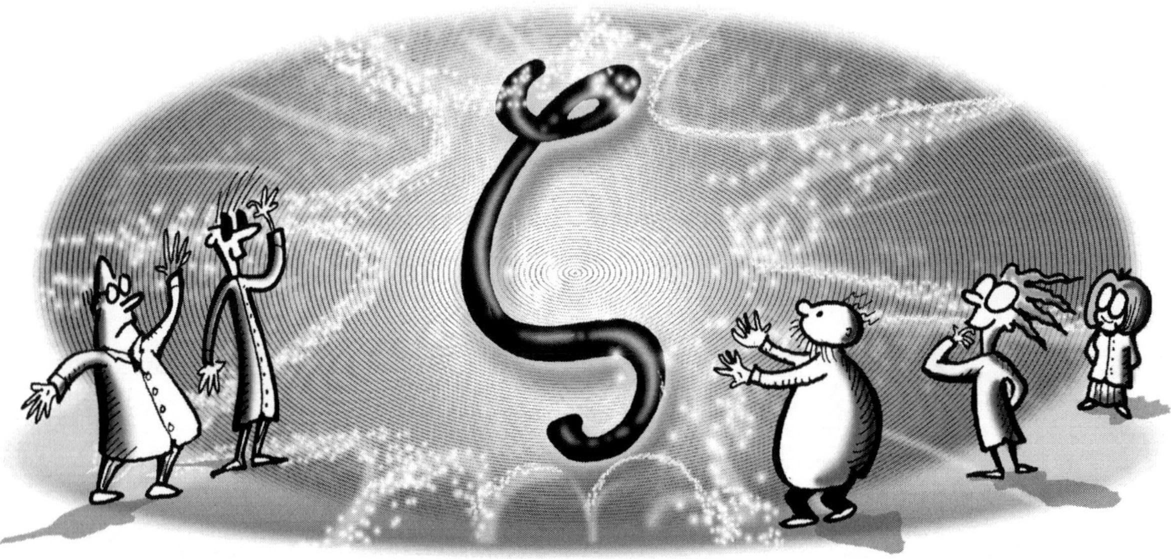

Chapter 22
Philosophical Intermission

If you *did* read Chapter 1 (you were given the option not to!), you might recall a long discussion about the distinction between "qualitative" and "quantitative" approaches to number. By page 33, though, we'd "put our quantitative heads on" and started to explore the prime numbers and their distribution. In this volume, the quantitative investigation has continued, leading to the Riemann zeta function and its zeros. But it's now time to take our quantitative heads *off* for a moment and have a look at what we've found from a more qualitative perspective.

THE "OTHER"

Although other mathematical objects have been described as "beautiful", *etc.* by mathematicians through the ages, I've found that there's a remarkable extent to which the zeta function inspires a sense of beauty and wonder in those people most familiar with it. It's been described as the most challenging, mysterious, fascinating and beautiful in all of mathematics.

> *"The zeta function is probably the most challenging and mysterious object of modern mathematics, in spite of its utter simplicity..."* (Martin C. Gutzwiller)[1]

> *"We may – paraphrasing the famous sentence of George Orwell – say that 'all mathematics is beautiful, yet some is more beautiful than the other.' But the most beautiful in all mathematics is the zeta function. There is no doubt about it."* (Krzysztof Maslanka)[2]

> *"...consider one of the most fascinating and glamorous functions of analysis, the Riemann zeta function..."* (Richard E. Bellman)[3]

The use of "glamorous" is interesting. The word "glamour" has changed its meaning over the centuries, from associations with witchcraft in the middle ages to the current usage, often associated with contemporary standards of female attractiveness. We'll never know quite how the quoted author intended to use it (he died in 1984), but in Volume 3 we'll see that much of the informal and nontechnical discussion of the mysteries surrounding the prime numbers and the zeta function involves language which evokes "the feminine". This is something I've discussed with a number of mathematically trained academics who are also familiar with the work of the psychologist Carl Jung. One suggested explanation which has come up in these discussions is that in these number theoretical mysteries, the (overwhelmingly male) mathematical community is encountering something truly *other*, something beyond its usual categories of experience.

In his fascinating 1989 book *Jerusalem: City of Mirrors*, Amos Elon observes the recurrence of feminine metaphors associated with the city of Jerusalem through the ages and concludes that "*Language was playing odd tricks on the patriarchal East. Public statements are often rooted in private dreams. When men are mystified, they often resort to the feminine gender.*" [4]

In 1997, Temple University's Warren D. Smith circulated a paper with the unorthodox title "Cruel and Unusual Behavior of the Riemann Zeta Function". It examines certain subtle properties of the zeta function and concludes:

> "*It's remarkable how the Riemann zeta function seems to be trying intentionally to deceive us!*" [5]

"Cruel" suggests a personification, while "intentionally" suggests an intent, in this case a deceitful one. The author clearly knew what the zeta function "is" (in the textbook sense) when he wrote these words and I don't think he meant to be anything more than slightly whimsical or poetic. But these words are perhaps further clues indicating a collective, unconscious tendency to relate to the zeta function in a certain way.

THE "RELIGIOUS"

In the cult film *2001: A Space Odyssey* (1968), a key role is played by the bewildering discovery of a perfectly smooth, geometric "megalith" buried beneath the surface of the Moon, beaming out a radio signal. There's a memorable scene in which a handful of scientists and authorities gather around the (excavated and upright) megalith in a sort of disoriented awe, presumably wondering "how did *that* get here?" Riemann's discovery of the nontrivial zeta zeros reminds me of this. The last century-and-a-half has seen a handful of experts gathering around the zeta function and its random looking "splatter" of nontrivial zeros in a similarly bewildered, awestruck way. And some relatively recent discoveries establishing various unexpected links between the zeta function and physics have served to intensify this feeling (we'll look at this in Volume 3).

Much has been written about the theological meaning behind *2001* – some people have interpreted the cinematic megalith as a religious symbol. Although I'm certainly not suggesting that it *should* be, it's not too difficult to imagine alternative histories (or parallel worlds) in which the Riemann zeta function has become an object of religious significance to a mathematically advanced culture of humans.

What am I trying to say by this?

First of all, the zeta function underlies our number system, around which we've structured our currently dominant model of reality (see Chapter 1). The precise sense in which it "exists" can be debated, but those most familiar with it are in no doubt that it is utterly fundamental to reality (or at least to all those aspects of reality where things can be counted and measured).

> "[It has been] *said that the zeros* [of the Riemann zeta function] *weren't real, nobody measured them. They are as real as anything you will measure in a laboratory – this has to be the way we look at the world.*" (P. Sarnak)[6]

Secondly, exploration of the zeta function is a gateway into what appears to be an endless chain of mysteries, through which human minds can contemplate the infinite, the eternal (which can inspire both awe and humility) *and* agree about it!

As well as the language of "the feminine" and "the other", we'll go on to encounter much surprisingly mystical, perhaps even "religious", sounding language evoked by mathematicians discussing the primes, zeta and related issues – words and phrases like *great mystery, wonder, exalted, majestic, miraculous, awed, impenetrable, profundity, secret source, Holy Grail* and, of course, *secrets of creation*.

If we entertain certain ideas from Chapter 1 where I suggested that science and economics may have become the "true religion" of the early 21st century, then the mathematical community could be viewed as a sort of priesthood. As number theory is the core (or "the queen"[7]) of mathematics, it could be further argued that number theorists are the "inner priesthood" of Western civilisation. And, at the very heart of number theory, we find a handful of little known scholars intensively studying the zeta function in almost unimaginable detail (each with his/her own private motivations).

THE "MYSTERIOUS"

I chose the name *The Mystery of the Prime Numbers* for Volume 1 because it had a populist ring to it, sounding like something accessible to people of all ages (something like a Sherlock Holmes story, perhaps)...and because it just sounds intriguing. "*What* mystery?" some people demand to know, associating prime numbers with boring, long-forgotten school lessons. But the word *mystery* has a deeper meaning than the one that people often think of (detective stories, or documentaries about lost civilisations and/or the paranormal), and this deeper meaning is relevant to our subject matter. In the early 14th century, "mystery" had a theological meaning: it meant *a religious truth obtained via divine revelation* or *the mystical presence of God*. This can be traced back to the Greek *mysterion/mysteria* (a secret document or rite) and further back to *mystes* (someone who has been initiated).

Think of a young person who has spent twenty years of her/his life studying maths and reached a point where, having earned a doctorate, s/he can begin to do original research (on the Riemann zeta function, for example). This twenty years of study would have centred on the reading and writing of symbols and numerals in strings and clusters, these becoming more and more indecipherable to the general populus the further the student proceeded in her/his studies. By the end of this time, s/he will have joined a relatively miniscule group of people who understand the details of the subject matter in question. This is arguably a form of initiation, comparable in some ways to an initiation into a priesthood, occult order or mystical sect.

Atheists and anti-mystical sceptics may forcefully argue that *no*, this is something very different, that it's the opposite of anything of a mystical or religious nature. And I agree that there *is* a fundamental difference: unlike the usual objects of religious devotion (holy scriptures, deities, saints, gurus, relics, sacraments, *etc.*), the Riemann zeta function is something which everyone concerned with *can agree about*. We're unlikely to see "schisms" in number theory due to factionalism or differing interpretations. Number theory is *entirely communal*: in as far as you can say this about anything, *it's true for everyone,* and anyone who wishes to make the effort can explore it and confirm its truths for her/himself. This is important and should be kept in mind, but it doesn't change the fact that to a truly "outside" observer (some sentient being from a very different world, perhaps) the processes of priestly initiation and studying to become a professional mathematician would appear very similar in many ways. It wouldn't be unreasonable to describe this trilogy as a kind of "initiation", hence its subject matter is (in the original sense) a *mystery*.

123

In Chapter 1, after considering the possibility that quantitative economics and science have replaced some of religion's traditional functions and observing that both of these activities are centrally underpinned by number, we looked at various connections between number and religion in its various forms. And now that you've read this far and got some feeling for the Riemann zeta function (lurking there behind the number system), you can perhaps see how it might inspire secular, professional mathematicians to use phrases like "*inexplicable secrets of creation*" and "*an arcane music and a secret harmony*".

Quantitative scientists and mathematicians (especially subatomic physicists and number theorists) have moved in to occupy the empty space left by theology and mysticism in the popular imagination, these having gradually lost their grip of the worldviews of the Western and Westernised masses. Unless overtly religious, Western adults are now tending to believe what (for example) Stephen Hawking says about the nature of things rather than (for example) the Pope or the Archbishop of Canterbury. As this historical shift occurs in the popular mind, it seems that (along with quantum mechanical discoveries and cosmological issues like the Big Bang), mathematics, the number system, the primes, the zeta function, *etc.* are becoming fertile ground for the blossoming of the sort of descriptive language which would previously have been reserved for religious or spiritual matters.

When a modern "scientific" society no longer believes in *religious truth obtained via divine revelation* (known as a *mystery* in the 14th century), does anything take its place? If anything, it would presumably be inspired moments, flashes of insight in which scientific truths are revealed to such as (arguably) Archimedes, Descartes, Newton, Einstein, Darwin, Crick and Watson or Hawking. Such discoveries in the physical sciences are very often superseded by more accurate or refined theories, though – they're almost always partial, incomplete or flawed in some way. The proof of a mathematical theorem, on the other hand, has an absolute, eternal nature *and*, rather than dealing with matter, it concerns a mysterious, invisible realm ("mathematical reality") which seems to impinge on this material world, despite no one really knowing "what" or "where" it "is".

So, if the nonreligious public were much more familiar with the greatest achievements of mathematicians throughout history (I'm not suggesting that they *should* be), then you could imagine that the greatest historical incidences of fruitful mathematical inspiration would occupy that place in the collective imagination which used to be reserved for religious *mysteries* (in the original sense). And it would seem reasonable that among the most profound of these mysteries would be those associated with the distribution of primes (the bones of the number system)...that is, with the Riemann zeta function.

THE "DIVINE"

The Indian mathematical prodigy Srinivasa Ramanujan provides us with the only claim I'm aware of involving (verifiable) higher mathematical insight being somehow obtained via religious or mystical channels, rather than purely cerebral ones. He was somehow able to produce highly elaborate formulas and results without the need for the usual chains of reasoning which lead up to them (these being what conventional mathematicians spend their days busying themselves with). Some of these results subsequently took other people years to construct proofs for. Ramanujan claimed that his Hindu family goddess Namagiri revealed mathematical truths to him in his dreams. And, interestingly, it was largely in analytic number theory that he worked, that branch of mathematics at the centre of which sits, enthroned, the Riemann zeta function.

I've come across unsubstantiated claims that Namagiri (more correctly "Lakshmi Namagiri") does have some association, among many others, with number and calculation, but I've been unable to verify this and suspect that it's just wishful thinking. However, it seems to me that *the zeta function itself*, personified, could quite easily join the vast pantheon of Hindu deities (there's at least one for just about every principle, virtue, concept, *etc.*) in the unlikely event that it were ever to be culturally absorbed[8]. The 19th century scientist Francis Galton once proclaimed of another iconic mathematical entity, the *Gaussian distribution* (most famous as "the bell curve"), that if the ancient Greeks had known about it, they would have personified and deified it [9]. This claim has attracted some controversy, but that concerns the application of the Gaussian distribution to physical reality, not the mathematics of the thing itself [10]. In its usual context involving the distribution of experimental data, the Gaussian has one foot in mathematical reality and one in physical reality. The zeta function, on the other hand, belongs wholly to mathematical reality. It requires no reference to the contents of the physical world, and I'm in no doubt that the ancient Greeks would have deified *it*, had they known about it (importantly, this *doesn't* mean that I think it "should" be deified!).

A curious webpage appeared some years ago, posted by a Swedish student, simply proclaiming (in large typeface) *"Feel the power of the Riemann zeta-function!"* with no other content [11]. More recently, an economics lecturer in Dublin produced a considerably longer and more nuanced document on the same topic, publishing an article online involving a holistic quantitative *and* qualitative approach to the issues surrounding the prime numbers and the zeta function [12]. In this, he attempts to discuss qualitative "meanings" of certain mathematical concepts, symbols and phenomena, relating them to various stages of personal spiritual evolution. Although the idea of such an approach is fascinating in theory, I find the author's grounding in number theory and psychology not strong enough to sustain his arguments. Still, it's of interest that this article – clearly a serious work of a serious mind – appeared at all.

Despite these scattered hints and clues, though, possible connections between analytic number theory and mysticism, religion or theology have not been discussed in any other literature that I'm aware of. But this is perhaps not surprising, considering that theologians and mystics have not been aware of the zeta function, and mathematicians generally believe that they have no need and/or time for theology or mysticism.

That Gaussian distribution (or bell curve) which so impressed Francis Galton turns out to be curiously relevant to the distribution of prime numbers, although he wouldn't have known that at the time. Most mathematicians don't even know it now. The ones who do are those familiar with an area of study called *probabilistic number theory*.

Probabilistic number theory is a fairly recent development: its first major result (sometimes called its "fundamental theorem"), the *Erdős–Kac Theorem*, didn't appear until 1940[13]. This is named after the famously prolific mathematician Paul Erdős (primarily a number theorist) and another mathematician, Mark Kac, who had a greater knowledge of probability theory than did Erdős (a 1939 lecture by Kac inspired Erdős to prove the theorem). Interestingly, this theorem was built directly on some earlier work that Srinavasa Ramanujan did with Godfrey Hardy (although I don't know if he claimed any help from Lakshmi Namagiri for that).

To explain the Erdős–Kac Theorem, I'll first have to explain the Gaussian distribution (for readers who haven't met it before). So we'll now put our quantitative heads back on and put aside the Riemann zeta function until the end of this chapter. This is something of a diversion from the main direction of our exploration, but it shows another (I think extremely important) face of the number system, one involving the philosophically subtle and problematic concept of *randomness*.

THE GAUSSIAN DISTRIBUTION

In decades past, the Gaussian distribution was known as the "Normal Law" and also as the "Law of Frequency of Errors". It's now most commonly known in terms of the "bell curve". The extent to which it's applied in the social sciences and economics (both of which are closely tied to human behaviour) is starting to come under serious critical scrutiny in the early 21st century, but here we'll look at an uncontroversial example where it could reasonably be applied in the (nonhuman) physical world.

Suppose you find a beach which is covered in smooth pebbles, all roughly of the same size, but (of course) each one slightly different. You gather a large number of these and weigh them individually. You then make a "bar graph" showing how many pebbles fall into each of a series of "weight brackets" (for example, 0–0.5g, 0.5–1g, 1–1.5g, 1.5–2g, 2–2.5g, *etc.*).

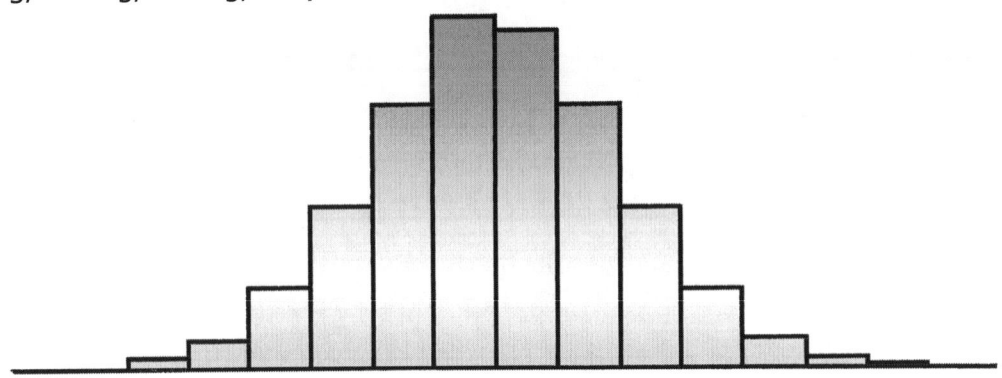

The base of each block represents one of the "weight brackets" and the block's height represents how many pebbles have been found in that weight bracket.

Next, you gather an even larger amount of pebbles and weigh them all, then construct the same kind of graph, but with many more "weight brackets" (each covering a narrower range than those used the first time – say 0–0.25g, 0.25–0.5g, 0.5–0.75g, 0.75–1g, 1–1.25g, *etc.*).

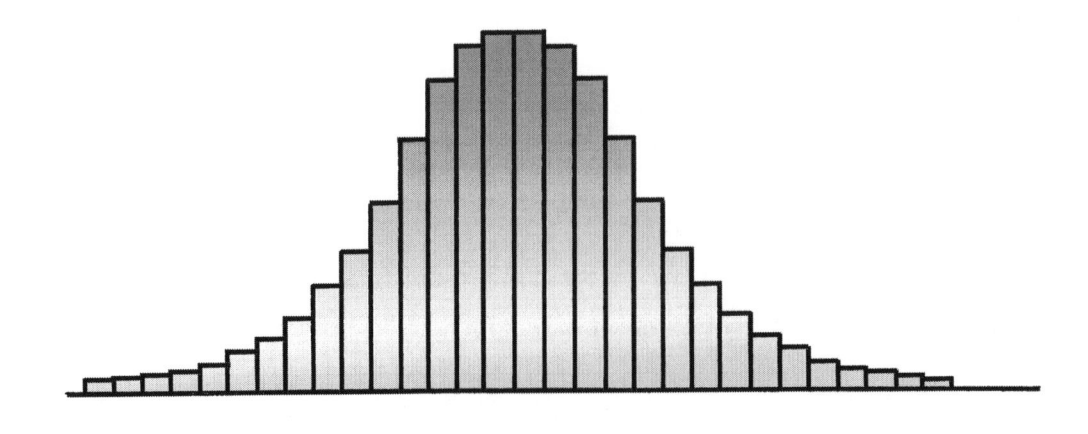

If you keep doing this (larger numbers of pebbles, narrower weight brackets) and you find that your bar graph (also called a *histogram*) starts to look more and more like this *bell curve*...

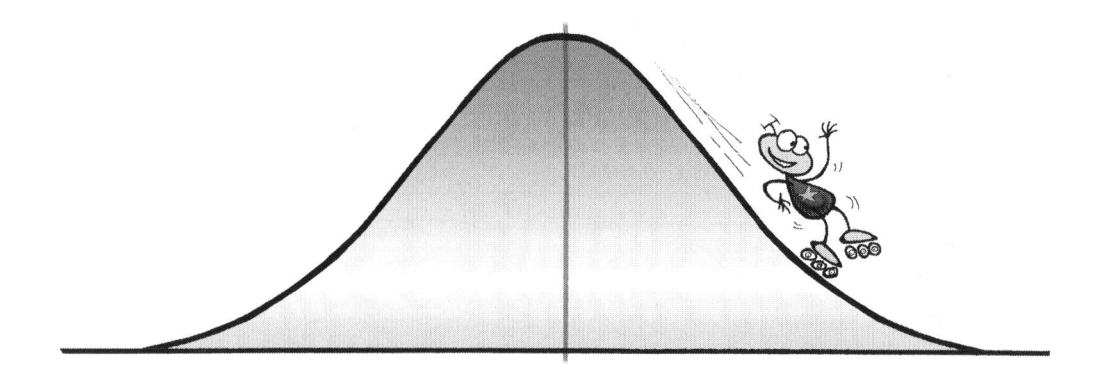

...then we say that the pebble weights are *distributed according to the Gaussian distribution*. *Many* things that we can identify, name, count and in some way measure (not just weigh) seem to be distributed like this, whether in the nonhuman worlds of botany, zoology, seismology, astronomy, *etc.*, or in human culture, involving sports results, economics, word and letter frequencies in printed literature, *etc*. Some people have argued that the extent of this (the so called "ubiquity of the Gaussian") is greatly overstated, but everyone agrees that it shows up a *lot* and in a dizzyingly diverse range of measurable phenomena from every realm of experience.

And the bell curve isn't just a vaguely bell-shaped curve, it's the graph of a precisely defined mathematical function:

$$\frac{1}{\sqrt{2\pi}} e^{-\frac{x^2}{2}}$$

129

Nassim Nicolas Taleb, one of the most outspoken critics of the overuse of the Gaussian distribution has pointed out that you never encounter this function (that is, a *true bell curve*) in the physical world[14]. You'll only ever find an *approximate histogram*. Because however many pebbles you weigh, star brightnesses you measure, texts you analyse, people you interview, earthquakes you measure, temperatures you record... you're still inevitably working with a finite amount of data. So your histogram will still be made of narrow vertical bars with flat tops. This can become as impressively close as you like, but it will never actually *be* a smooth curve (imagine repeated attempts to construct a bell curve out of wooden blocks, each time using thinner blocks).

THE ERDŐS–KAC THEOREM

So the bell curve is a Platonic ideal, with a kind of "archetypal" existence. We see "shadows" that it casts into the world, but we never meet one *in* the world.

Except that we *do*! If we extend our idea of the "world" from just the outer world of matter to also include the inner world of mind, then we quickly find the source of *an infinite supply of data*. As I've pointed out, we'll never be able to collect more than a finite number of pebbles or see more than a finite number of stars through our telescope. *All physical data is finite.* But the *number system* (which we can gain access to via our inner world) supplies us with an *infinite* source of data.

Obviously, the supply of counting numbers is infinite, but that's not really what I mean. By definition, counting numbers follow each other in a predictable sequence, gradually getting larger – there's no irregularity, patterning, or substantial "information content" to be found there. We're looking for a source of data which involves (or at least appears to involve) *randomness*. And so we look to the primes. As Oxford University mathematician Timothy Gowers has observed, "*Although the prime numbers are rigidly determined, they somehow feel like experimental data.*"[15]

It's not to the primes themselves, but to the *prime factorisations of the counting numbers* which we're going to look. You might remember this image from Chapter 4:

This was intended to give you a flavour of the "randomness" associated with the primes and their interaction with the number system as a whole. Now, as you can imagine, if you were to walk along the number line "cracking open" each counting number to find its prime factorisation, there would be a slow, general tendency for the amount of primes involved in the factorisations to get larger. There would still be significant variation since you'll continue to meet, and crack open, prime numbers (they have the smallest possible amount of factors, that is, one). Just as 31 (a single-prime factorisation) sits next to 32 (five factors), we can expect to see even bigger jumps from number to number. But there's an *overall tendency* for factorisation sizes – the amount of spheres in each cluster – to grow. This is what you'd expect: bigger numbers, bigger clusters, bigger factorisation sizes. We're going to "measure" these in a very precise way which I shall next describe.

The theorem of Hardy and Ramanujan which I mentioned a few pages back has a similar character to the PNT. Recall that the PNT says (roughly) that the number of primes less than a large number is approximately that number divided by its logarithm. The Hardy–Ramanujan theorem states (roughly) that the number of prime factors of a large number is *approximately the logarithm of the logarithm of that number*. This "approximately" is in a similar sense as in the PNT. In this case, it involves the proportion of counting numbers whose factorisation sizes deviate by more than a certain amount from this "log log" approximation, and the way that this "misbehaving" proportion can dwindle as close to zero as you like if you're prepared to travel far enough along the number line. The precise statement of the theorem need not concern us, the "logarithm of the logarithm of the number" is all we need to take from this.

I should clarify that this theorem (and the Erdős–Kac Theorem too) concerns *distinct* prime factors, in the sense that 35 = 5 × 7 and 24 = 2 × 2 × 2 × 3 both have two distinct prime factors, 125 = 5 × 5 × 5 has one distinct prime factor and 30030 = 2 × 3 × 5 × 7 × 11 × 13 has six. The *total* number of factors ("cluster size") can obviously be much larger than the number of *distinct* factors for any given number.

To explain that "logarithm of the logarithm of the number" visually, you'd have to count the coils of a spiral inside a circle whose radius was that number, then make a circle with radius size equal to the number of coils you just counted…then count coils again – we won't attempt to illustrate that here, as you should have got the idea of logarithms by the time you finished reading Volume 1.

If, at each counting number, we (1) find its factorisation, (2) count the number of distinct prime factors involved and (3) subtract the logarithm of the logarithm of the counting number from the number of distinct prime factors, then we sometimes get negative numbers (smaller-than-average amount of distinct factors) and sometimes positive numbers (larger-than-average amount). The final step is to divide each of these numbers by the *square root* of the logarithm of the logarithm. To get a visual sense of square roots (which have hardly been mentioned until now), you can consider this picture, illustrating the square root of 6.9 by means of an equiangular spiral:

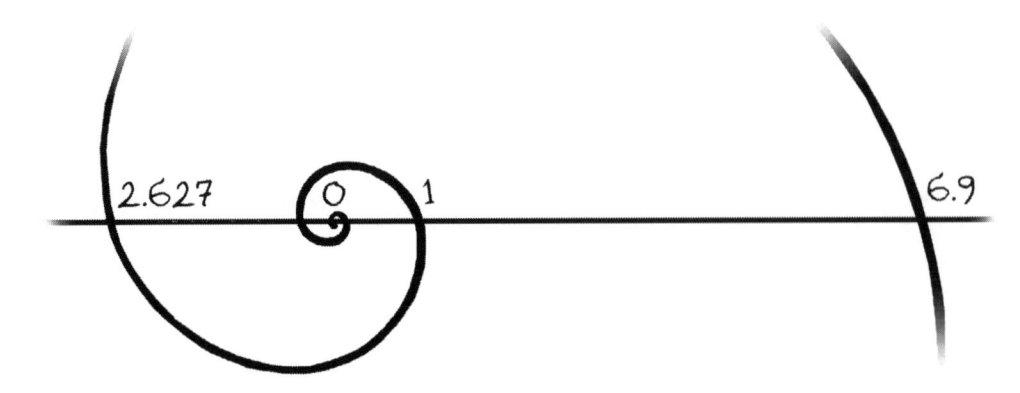

Here we see the spiral with base 6.9. Its first crossing of the *negative* horizontal axis, travelling clockwise from 1, is 2.627…, which gives 6.9 when multiplied by itself.

So, to sum up, we take the number-of-distinct-prime-factors at each counting number, subtract something from it, divide the result by the square root of that same something... and what we're left with is our "measurement", an item of data. It's an irrational number which could be positive or negative. Large negative measurements would indicate a considerably-smaller-than-expected number of distinct prime factors. Large positive measurements would indicate a considerably larger-than-expected number. Small negative measurements would indicate a slightly-smaller-than-expected number, and small positive measurements would indicate a slightly-larger-than-expected one.

If we were to do this for the first 10 000 or 100 million or 50 trillion (or whatever) counting numbers and in each case collect all these "measurements" and make a histogram with appropriately narrow "size brackets", then the Erdős–Kac Theorem guarantees (they *proved* this) that the histograms will start to get closer and closer to a perfect bell curve (as close as you require), the farther down the number line we go.

THE NOTION OF A "LIMIT"

In mathematical analysis, you very often encounter the phrase "let n tend to infinity". Here "n" is actually n, a "variable" which represents an arbitrary counting number. These "let n tend to infinity" statements generally involve certain things getting closer and closer to certain other things as n gets larger. If we let n run off towards *infinity*, then these things become "infinitely close", that is, they effectively become equal. This is the basic idea of what mathematicians call a *limit*.

The simplest example is "$1/n$". The limit of this quantity as n tends to infinity is 0 as we get the sequence 1, 1/2, 1/3, 1/4, 1/5,... None of the fractions in this sequence will actually *be* 0, but they will eventually come as close as you choose to specify.

This is the language that the Erdős–Kac Theorem is framed in. As n goes to infinity, the histograms eventually become as close to a bell curve as you choose to specify. So the limit of our sequence of histograms is the Gaussian distribution.

A rarely discussed philosophical problem with probability theory is the way that certain key theoretical results (the so called *Central Limit Theorem* and also the *Law of Large Numbers*) involve this letting *n* tend to infinity. These fixtures of probability theory are ultimately founded on the notion of statistical data, based on the human experience of measurable events in the real world. The whole of the theory of probability and statistics rests on the idea of (comparable) "events". It refers to the physical world, and in this way distinguishes itself from other areas of mathematical theory. And in the real world of measurable objects and events, you *can't* let *n* tend to infinity. There's only a finite supply of whatever it is you're working with (and there's only a finite amount of time to measure it in).

As Godfrey Hardy and John Littlewood once observed, "Probability *is not a notion of pure mathematics, but of philosophy or physics.*"[16] Something isn't quite right here: a pure mathematical notion (letting *n* tend to infinity) is being applied in a realm where it doesn't properly belong, the messy physical world of objects, events, measurements and *finite amounts of data*.

However, in 1940, centuries after the beginnings of probability and statistics (which originated in a sprawling, ancient accumulation of gambling techniques and tricks), a truly infinite data set – the data derived from the prime factorisations – was viewed through the lens of probabilistic analysis *and shown definitively to be distributed in a precisely Gaussian way*. This means an *exact* bell curve. Now we *can* let *n* tend to infinity, because unlike pebbles, stars, grains of sand or snowflakes, there's an infinite supply of counting numbers for *n* to represent.

So we'll never find a perfect bell curve in "real world" data, but we *can* find one in the number system. In fact, it turns out that there are many different "central limit theorems" in number theory: in each case, a diversity of things (always precisely defined and of which there are infinitely many) are measured, leading to precise Gaussian distributions[17]. The "splatter" of the Riemann zeros, for example, has been shown to display what is known as "Gaussianity" when viewed in the appropriate way. The inherent "randomness" of the number system, visible in many of its facets, allows this probabilistic concept to find perfect, precise, pure realisation ... and nowhere else can it find that, all instances in the physical world being imperfect realisations (however slightly).

Even the notion of "statistical independence", which is at the very core of probability theory, can only *truly* be realised in the context of number theory. Theoretically, probabilities associated with consecutive rolls of a die should not affect each other. But because modern scientists are now aware of the wavelike nature of matter and of the inseparable nature of sp;ace-time and its contents, they can no longer deny that there's always going to be *some* degree of "interference" (however miniscule) between any pair of rolls. The probabilities associated with the "events" of an arbitrary (and appropriately large) counting number being divisible by two different primes, though, are *truly* independent. It's as if, in the probabilistic number theory setting, the notion of statistical independence can be realised in some Platonic or archetypal way.

A BUDDHIST PERSPECTIVE

Since first learning about the Erdős–Kac Theorem (and the whole family of related "central limit theorems" which identify perfect Gaussian distributions in other aspects of the number system), I've not been able to let go of the feeling that there's something of unacknowledged – and considerable – importance about it. This is partly because I became convinced some years earlier, and for no reasons which I can clearly recall, that there *must* be a connection between the prime numbers and the Gaussian distribution – I tried in vain to find such a connection before eventually learning (vindicated, yet still amazed) of the Erdős–Kac theorem. But it's also because of another realisation concerning the application of the Gaussian distribution to the outer world of matter and "events".

This can be related to the early Buddhist concept of *jñeyāvarana* (or "veil of fixed ideas"). Philosophers from this tradition have explored the idea that the "outer world" which we tend to believe we're interacting with is *not the world itself*, but rather a sort of veil that lies between our minds and the world "as it truly is". We effectively cover the world-as-it-is with a mesh of fixed concepts, opinions and (most importantly) categories of named things. This projection of categories onto the world is an unconscious layer of culture – it's never really acknowledged or discussed (except by a few philosophers), it just gets passed down from generation to generation.

It's in this adult-to-child transmission of categories where counting can (and generally does) enter our experience. In most cultures, very young children learn to recognise and name things which belong to certain categories and, at the same time, they learn to count (to the extent that their culture emphasises counting) – these two activities go hand in hand. As I pointed out in Chapter 1, to count, *you need a category of thing-to-count*. This process of actively exposing a child to our most basic cultural contents (categories of named objects, and numbers), despite being a perfectly natural response to the world as we find it, is basically the act of constructing the initial layer of *jñeyāvarana* in that child's mind, what the early Buddhist philosophers called *prapañca*, usually translated as the "multiplicity of named things". It's not at all clear to me what other options are available, and social and cultural forces make this kind of enculturation almost inevitable, but I think it's important that we at least acknowledge what we're doing.

Our categories are not "in" the world, nor are they entirely in our minds, they're (and this is unavoidably vague) *somewhere in between*. So, although the category "pebbles on this particular beach" seems like a perfectly natural one, it's dependent on definitions and assumptions which have nothing to do with the world in its raw state. For suppose a large number of spherical plastic objects washed up on the beach, would we gather and weigh them along with the pebbles and include them in our data gathering? No, of course not, because *we know they're not pebbles*. If we were dealing with a robot helper or an extraterrestrial employee, though, we'd have to come up with precise definition of "pebble" in order to avoid non-pebbles being included in our survey. You might think that this would be fairly easy to do. But then what if a storm washed a number of noticeably larger stones up onto the beach – would you redefine "pebble"? If so, where does "pebble" begin and end? It doesn't, you see, it's just an arbitrary human imposition on the "raw" world which just happens to suit our purposes at any given time. And it's not *in* the world. It's something we project *onto* the world. It's part of the *jñeyāvarana*.

Continuing to think along these lines, we must conclude that...

any (approximate) bell curves found through experimental science and the gathering of data from the physical world are not "in" the world – all such things are in the "veil"

...while remembering that...

they're never going to be perfect bell curves – the only *perfect ones are found hidden within the number system.*

And how does the number system relate to this "veil of fixed ideas"? This is a big question, one we'll return to in Volume 3. But it's now time to put such metaphysical matters aside and return to the Riemann zeta zeros and a more straightforwardly mathematical mystery in which they play the central role.

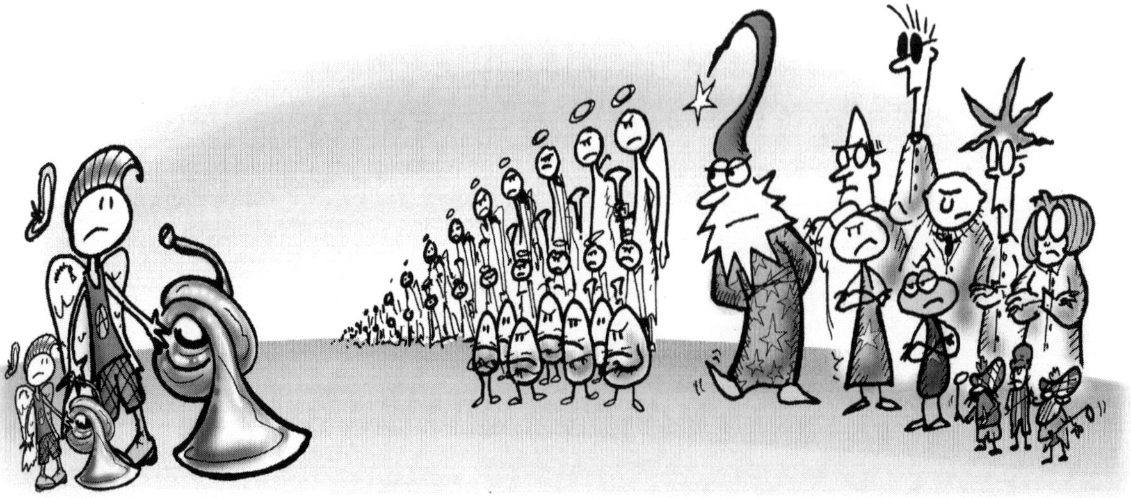

Chapter 23
the Riemann Hypothesis

Riemann's suggestion in his classic 1859 paper that all of the nontrivial zeros of his zeta function should lie on the critical line has become known as the *Riemann Hypothesis* (we'll abbreviate this as "RH").

This is all Riemann's paper had to say about his hypothesis, after proposing it:

> *"Certainly, it would be desirable to have a rigorous proof of this proposition; however, I have temporarily put aside my search for this after some brief and unsuccessful attempts, since it appears to be unnecessary for the immediate goal of my investigation."* [1]

Remember, Riemann's immediate goal at this time would have been to prove the PNT, which, he'd realised, amounted to proving that there aren't any zeros on the boundary of the critical strip (the vertical lines through 0 and 1).

As I write this (the summer of 2011), we still don't know whether or not the RH is true, and the significance it has acquired in mathematics in the intervening 152 years can hardly be overstated.

More and more areas of mathematics have been shown to be linked to it in some way. Beyond this, quite a lot of mathematical theory has been put forward and shown to be true *provided the RH is true* (so the validity of that work stands or falls with the RH).

Also, as I've mentioned (and as we'll see in Volume 3), the zeta function has been showing up unexpectedly in *physics* in recent years, causing much surprise among mathematicians and physicists, so the desire to understand its behaviour has become more heightened than ever. And the RH is central to any such understanding.

David Hilbert, a giant of late 19th and early 20th century mathematics, once stated in an interview that he believed the Riemann Hypothesis was the most important problem "not only in mathematics, but *absolutely [the] most important*"[2]. He also allegedly said that if he were to wake from a thousand-year sleep, the first thing he'd ask would be "Is the Riemann hypothesis established yet?"

Having got through Volume 1 and this far into Volume 2, you may agree that the Riemann Hypothesis – a proposal about some points lining up in a weird, abstract "number plane" – is an interesting matter... but "absolutely the most important" problem in the world?? Clearly, not everyone would agree with this (not even every analytic number theorist), but I'm going to try to convey what I feel Hilbert was implying as we proceed further into the heart of the number system.

For now, here's what some other mathematicians have had to say about it:

> "...*the Riemann hypothesis remains one of the outstanding challenges of mathematics, a prize which has tantalized and eluded some of the most brilliant mathematicians of this century.*" (Richard E. Bellman, 1961)[3]

"Whoever proves or disproves it will cover himself with glory..." (Eric T. Bell, 1937)[4]

"I am firmly convinced that the most important unsolved problem in mathematics today is the truth or falsity of a conjecture about the zeros of the zeta function, which was first made by Riemann himself..." (Enrico Bombieri, 1992)[5]

"Ask any professional mathematician what the single most important open problem in the entire field is and you are almost certain to receive the answer 'the Riemann Hypothesis'." (Keith Devlin, 1999)[6]

"So if you could be the Devil and offer a mathematician to sell his soul for the proof of one theorem – what theorem would most mathematicians ask for? I think it would be the Riemann Hypothesis." (Hugh Montgomery, quoted by K. Sabbagh, 2003)[7]

"The Riemann Hypothesis is a precise statement, and in one sense what it means is clear, but what it's connected with, what it implies, where it comes from, can be very unobvious." (Martin Huxley, quoted by K. Sabbagh, 2003)[8]

"The Riemann Hypothesis is the central problem and it implies many, many things. One thing that makes it rather unusual in mathematics today is that there must be over five hundred papers – somebody should go and count – which start 'Assume the Riemann Hypothesis', and then the conclusion is fantastic. And those [conclusions] would then become theorems...With this one solution you would have proven five hundred theorems or more at once." (Peter Sarnak, quoted by K. Sabbagh, 2003)[9]

"If it's not true, then the world is a very different place. The whole structure of integers and prime numbers would be very different to what we could imagine. In a way, it would be more interesting if it were false, but it would be a disaster because we've built so much round assuming its truth." (P. Sarnak, quoted by K. Sabbagh, 2003)[10]

"A solution to the Riemann Hypothesis offers the prospect of charting the misty waters of the vast ocean of numbers. It represents just a beginning in our understanding of Nature's numbers." (Marcus du Sautoy, 2003)[11]

To summarise, the Riemann Hypothesis is supremely important within the world of mathematics and possibly even in a much wider context. Its persistent resistance to proof has resulted in fame and fortune becoming guaranteed to anyone who ever succeeds in proving it. Beyond this, the inherently intriguing and fundamental nature of the problem, as well as the fact that so much of mathematics has either been related to the RH or is built on the assumption of its truth, has driven some people to obsess over the problem. As humanity's knowledge concerning the number system currently stands, the RH is the central mystery. To pursue a proof of this hypothesis is to seek a key to unlock some of the number system's most closely guarded secrets.

One of the reasons for all the interest in the RH was discussed in the last chapter. If it's true, then the "PNT error" is as small as possible, so the prime numbers stay as close to their "average behaviour" as they possibly can (while still being constrained by the fact that they must produce the whole set of counting numbers when we take all possible combinations of them as factors). This is sometimes described in terms of the primes being "as well behaved as possible".

If the RH is false, then the PNT error is *not* as small as possible, the primes don't stay as close to their average behaviour as possible, and so they're not "as well behaved as possible". This would be caused by certain spiral waves from the infinite collection that "composes" the deviation having amplitude growth rates greater than ½ due to pairs of nontrivial zeros lying off the critical line (to the right). Recall that the horizontal position (real part) of a pair of vertically mirrored nontrivial zeros equals the amplitude growth rate of the associated pair of spiral waves.

As Enrico Bombieri has explained:

> *"Even a single exception to Riemann's conjecture would have enormously strange consequences for the distribution of prime numbers...If the Riemann hypothesis turns out to be false, there will be huge oscillations in the distribution of primes. In an orchestra, that would be like one loud instrument that drowns out the others – an aesthetically distasteful situation."* [12]

These "huge oscillations" would be the direct result of the spiral waves with amplitude growth rates bigger than ½. On another occasion, he said:

> *"The failure of the Riemann hypothesis would create havoc in the distribution of prime numbers. This fact alone singles out the Riemann hypothesis as the main open question of prime number theory."* [13]

All numerical evidence found as of 2011 (widely thought of as being "a lot") supports the view that the RH is true. And its being true does seem the more "balanced" of the two possibilities (truth or falsehood), as well as the more aesthetically pleasing:

> *"Mother Nature has such beautiful harmonies, so you couldn't say that something like that is false."* (Henryk Iwaniec, quoted in K. Sabbagh, 2003) [14]

> *"I would like the Riemann Hypothesis to be true, like any decent mathematician, because it's a thing of beauty, a thing of elegance..."* (A. Ivić, quoted in K. Sabbagh, 2003) [15]

Another aspect of this "well behavedness" (or not) is related to the "predictability" of the primes. If there were a spiral wave involved with an amplitude growth rate larger than ½, then it would "slosh the primeness around" in *huge* wavy fluctuations. If you were to look far enough down the line, then its peaks and troughs would eventually get as many times higher/lower than those of the "real part ½" waves as you care to imagine, which roughly means *that many times more or less primeness*.

Because those huge bumps above/below the horizontal axis mean "a lot more primeness here"/"a lot less primeness here"...

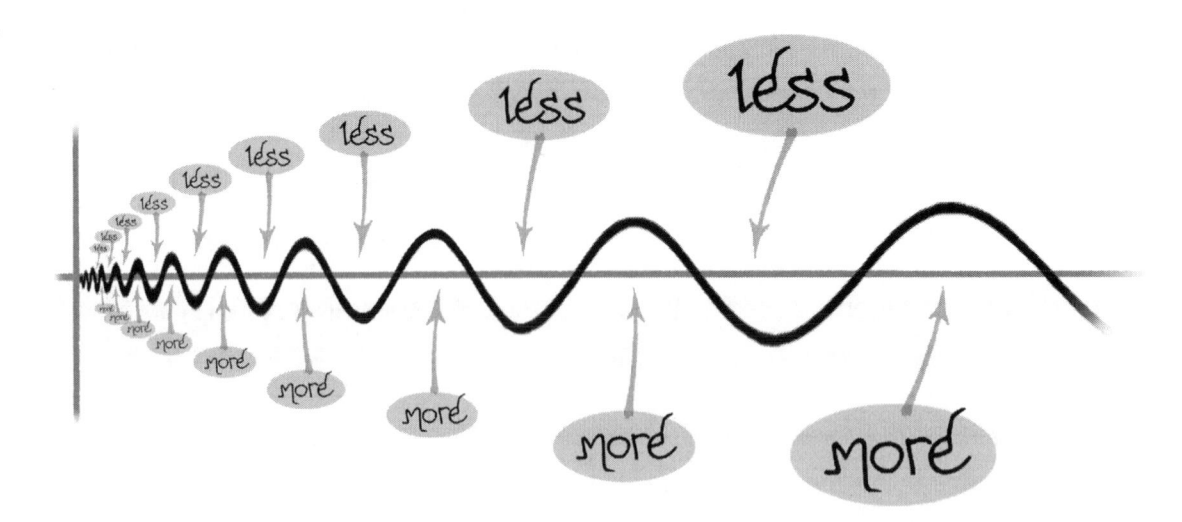

...if you were placing a bet on where to find a prime, you could gain an edge over your competitors with precise knowledge of a wave such as this. If the RH is true, then there are no such waves, therefore no such huge fluctuations, meaning "better behaviour" as regards predictability – in this situation, that entails the primes being, in some sense, as *unpredictable* as possible. This ties in with the issue of the "randomness" of the primes, an important matter to which we'll return in the next chapter.

PROGRESS ON THE RIEMANN HYPOTHESIS

So, how close are we to proving the RH?

I mentioned in the last chapter that Hardy and Littlewood proved that infinitely many nontrivial zeros lie on the critical line and that, decades later, Brian Conrey proved that at least 40% of them do. But neither of these results brings us an iota closer to the truth of the RH, any more than does the impressive (to early 21st century humans) numerical evidence produced by ZetaGrid or Andrew Odlyzko.

There's also been a result proved showing that "most nontrivial zeros are *very near* the critical line", in the following precise sense:

At each height above the horizontal axis, divide 8 by the logarithm of that number and then put a mark that many units either side of the critical line. Here we see a few examples:

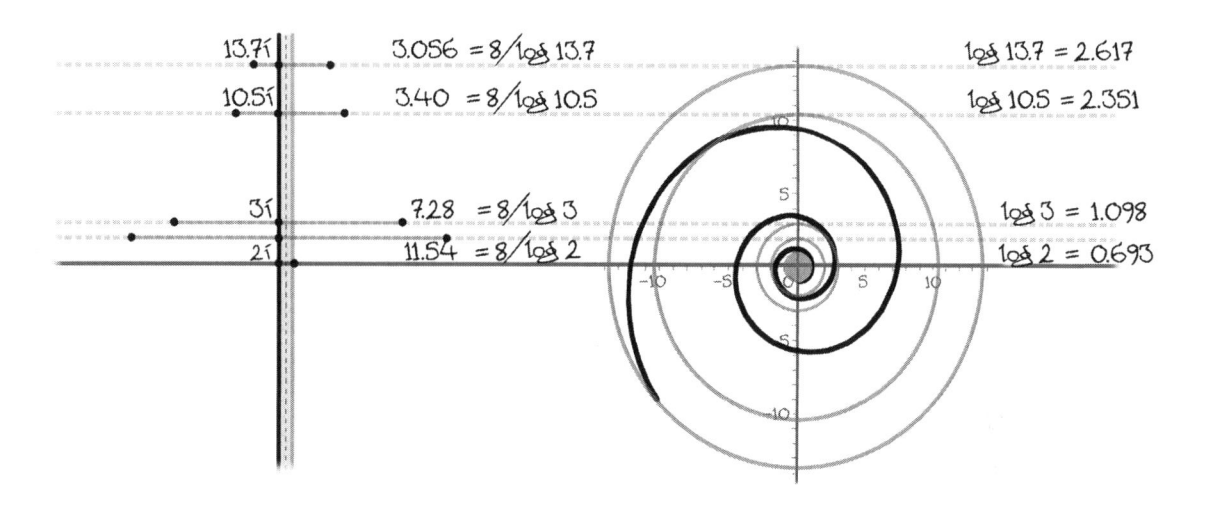

$13.7i$ $3.056 = 8/\log 13.7$ $\log 13.7 = 2.617$
$10.5i$ $3.40 = 8/\log 10.5$ $\log 10.5 = 2.351$

$3i$ $7.28 = 8/\log 3$ $\log 3 = 1.098$
$2i$ $11.54 = 8/\log 2$ $\log 2 = 0.693$

Doing this for all heights and then taking the entire image which is produced and reflecting it across the horizontal axis, you get this:

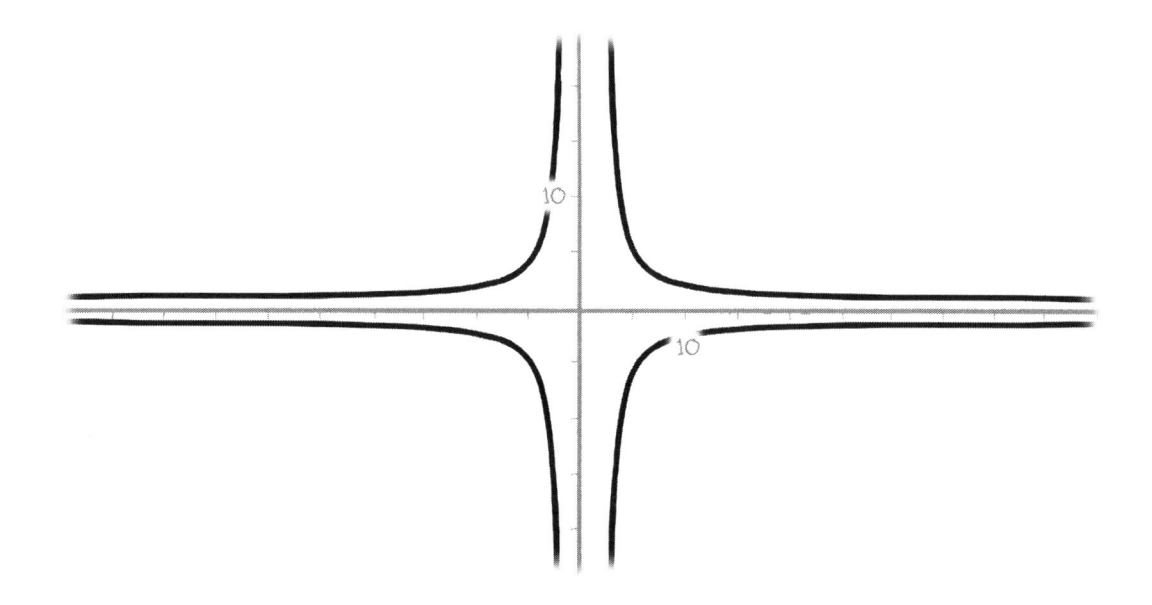

It has been proved that more than 99% of the nontrivial zeros lie inside this region. As you can see, it gets increasingly narrow as you travel up or down the vertical axis, eventually becoming "as narrow as you like" if you're prepared to travel far enough. Note that we have to travel up to almost 9 000 000 before this region narrows down to within the critical strip itself, and up to almost 80 000 000 000 000 before it's down to half of the strip's width. These might seem like unhelpfully large numbers, but remember that however high we climb up the critical strip, the infinite majority of nontrivial zeros with positive heights *will always be above us*.

The best "evidence" for the RH, if it can be called that, involves mathematics far beyond the scope of this book. But really, very little progress has been made.

ADDITION, MULTIPLICATION AND THE RIEMANN HYPOTHESIS

During the 20th century, many *reformulations* of the Riemann Hypothesis appeared, across many areas of mathematics (and even physics). A "reformulation" of the RH is *some other unproven proposition whose truth has been proved to be equivalent to the truth of the RH*. Despite the remarkable diversity of these reformulations (some of which we'll see in the next chapter), their proliferation is not as surprising as it might first seem if we recall that the RH concerns the prime numbers and the relationship between addition and multiplication, that is, the very foundation of the system of counting numbers, which underlies all of mathematics.

> *"The Riemann Hypothesis is the most basic connection between addition and multiplication that there is, so I think of it in the simplest terms as something really basic that we don't understand about the link between addition and multiplication."*
> (Brian Conrey, quoted in K. Sabbagh, 2003)[16]

> *"It is probably the most basic problem in mathematics, in the sense that it is the intertwining of addition and multiplication. It's a gaping hole in our understanding…"*
> (Alain Connes, quoted in K. Sabbagh, 2003)[17]

That the RH concerns the prime numbers should be clear to you by now. To get a better picture of how it relates specifically to the relationship between addition and multiplication, we'll return to some ideas from Volume 1.

Remember that there are two ways in which we can build $\mathbb{N}$: (1) the *additive* (Peano axioms) approach where we start with 1 and keep adding 1 to it, in this way producing the sequence 1, 2, 3, 4, 5,..., and (2) the *multiplicative* approach where we start with the primes and multiply them together in all finite combinations: 2, 3, 2×2, 5, 2×3, 7, 2×2×2, 3×3, 2×5, 11, 2×2×3, 13, 2×7, 3×5, 2×2×2×2, 17,... while thinking of 1 as the special combination involving "no primes being multiplied".

The seeming "awkwardness" of the prime numbers is a reflection of the fact that addition and multiplication don't neatly fit together – there's a sort of friction or tension involved. To put this in perspective, we'll briefly reconsider a visualisation from Chapter 6.

We're back in the infinitely huge, dark room full of clusters of glowing spheres, each sphere marked with a prime and such that *every* possible (finite) combination of primes is represented, and represented only once. The problem (we shall imagine) is that we're from a culture which uses a different set of number symbols, so we don't recognise the numerals "0", "1", "2", "3", "4", "5", "6", "7", "8" or "9". We understand about addition, multiplication, prime numbers, *etc.*, but these symbols on the spheres are just alien squiggles to us. So, we decide that *it doesn't actually matter* about the symbolic representation. We start with 1 (no cluster needed), then consider the next number, 1+1. We know this must be prime because there are no numbers bigger than 1 which you can combine to get it – the only number less than it is 1. So it can't possibly have any prime factors (other than itself).

So we pick a single-sphere cluster and decide that whatever its manufacturer intended, *for us* it's going to represent the first prime. Whether the symbol on it says "2" or not doesn't matter, as long as we stick with the notion that this represents the first prime. We can then take all clusters built solely from spheres marked with this symbol. These will correspond to $2 \times 2 = 4$, $2 \times 2 \times 2 = 8$, $2 \times 2 \times 2 \times 2 = 16$, *etc.* By counting out $1 + 1$ twice, then that twice, then that, *etc.* on the number line, we can correctly locate all of these "powers of 2" clusters.

We now have a hole between 2 and 2×2. It can't be arrived at by adding together copies of numbers bigger than 1, as there's only 2, and $2 + 2$ is already too big. Therefore it must be a prime. So, we pick another single-sphere cluster (again, it doesn't matter which symbol appears on it) to represent the second prime, that is, 3. It and all of its "power clusters" fall into place on the line, and any clusters involving just 2's and 3's (6, 12, 18, 24, 36, 54, *etc.*) can eventually be located on the number line by patiently counting out groups of 2 and 3.

There's then a hole between 2 × 2 and 2 × 3, so you choose another arbitrary single-sphere cluster to represent 5, then locate all of the clusters involving just 2, 3 and 5 on the number line, and so on.

Obviously, every "grid point" on the number line will get covered in this way because whenever we encounter a hole, it's immediately taken to be the next prime (and then all the remaining clusters involving just the primes up to and including that one can be located on the number line by counting). But a consequence of the Fundamental Theorem of Arithmetic – *and we really don't understand this to its core yet* – is that we *never* count out the location of a new cluster on the number line and find its position to be already occupied. This is the idea of *unique factorisation* – no two "prime recipes" can produce the same counting number. So these locations are always found empty. There's never any repetition or redundancy. And this goes on *forever*.

It's as if, via these addition- and multiplication-based approaches, the number system is "interacting with itself" in some infinitely complicated yet "infinitely perfect" way. And whenever numbers are invoked in human affairs, this remarkable situation is somehow there in the background.

By elaborating on this truth known as the Fundamental Theorem of Arithmetic, developing the idea of the Euler product formula as an "analytic" version of it and then extending this to $\mathbb{C}$ to find the Riemann zeta function (which expresses something even deeper about it), we end up faced with the enigma known as the Riemann Hypothesis.

The RH encapsulates the question of whether, once all the clusters in the room have been located on the number line (that would requite an eternity, of course), the single-sphere clusters distribute "as nicely as possible" among all the others. And although this "nice behaviour" is generally described in terms of "bounds" on the rate of growth of the primeness count deviation, there are other, equivalent, ways of describing it, as we shall next see.

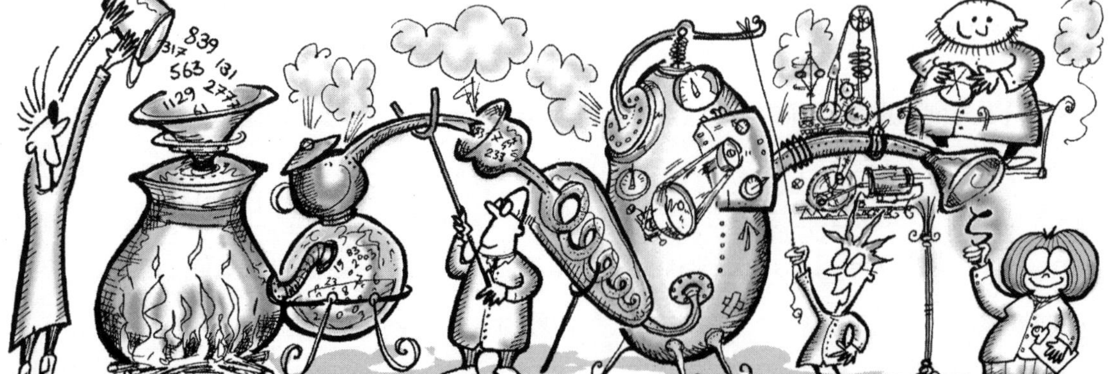
839
317
563
131
1129

chapter 24
reformulations of the Riemann Hypothesis

As I've suggested, with the Riemann Hypothesis being so central to the workings of the number system, it's not that surprising that it can be reformulated in so many ways, involving such a diversity of areas of mathematics and physics. Unfortunately, most of these reformulations involve mathematics that I'm not going to be able to explain in this book. But there are still a few that we can look at.

The first of these we're already familiar with. It concerns the rate of growth of the primeness count deviation. I've mentioned this in vague terms, but how do you actually measure the growth of something irregular like this? Despite its fluctuations, it's clearly growing overall, but it's not at all clear how you'd quantify that.

Mathematicians, particularly those working in the field of *analysis*, are often seeking to "bound" or "find bounds for" various functions. In order to get some sense of this visually, we'll work with graphs.

We'll need to create some special graphs for this purpose. I'll call them "power curves", which isn't what mathematicians would generally call them, although they'd understand this use of terminology. To do this, we first need a "power". This can be any number — we'll choose 1.637.

Now, for any given point on the number line, we imagine drawing the logarithmic spiral with that base, in other words, the logarithmic spiral that passes through 1 and then through that point as its next crossing of the positive half of the number line.

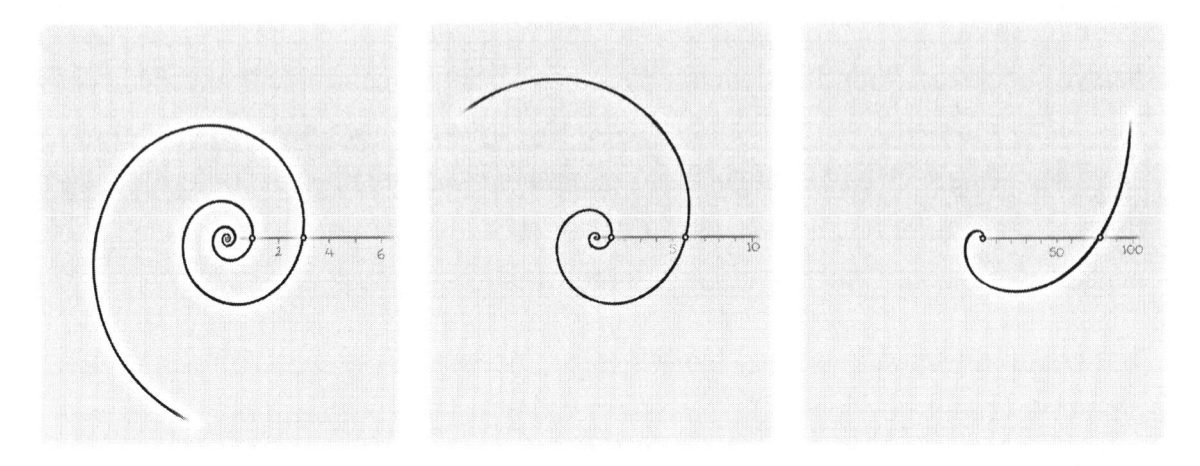

For three examples of points on the number line, 3, 5.7 and 79, we see the spiral with that base.

Now, starting at 1, travel 1.637 (the "power" we decided on) coils anticlockwise and then measure your distance from the centre:

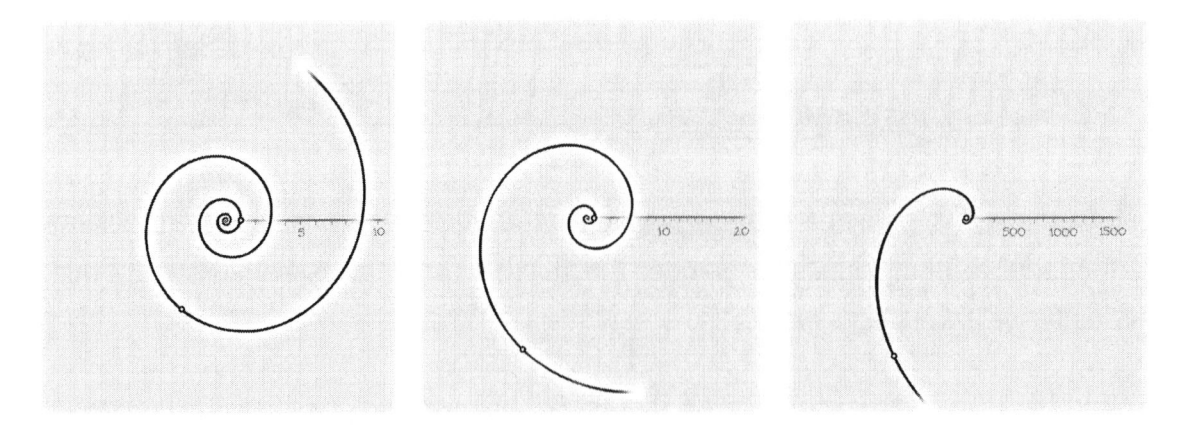

For our examples, the distances from the centre are approximately 6.040, 17.272 and 1277.660.

Now mark a point that high above the point you're at on the number line. In the leftmost panel below, we see our first two examples along with a number of others.

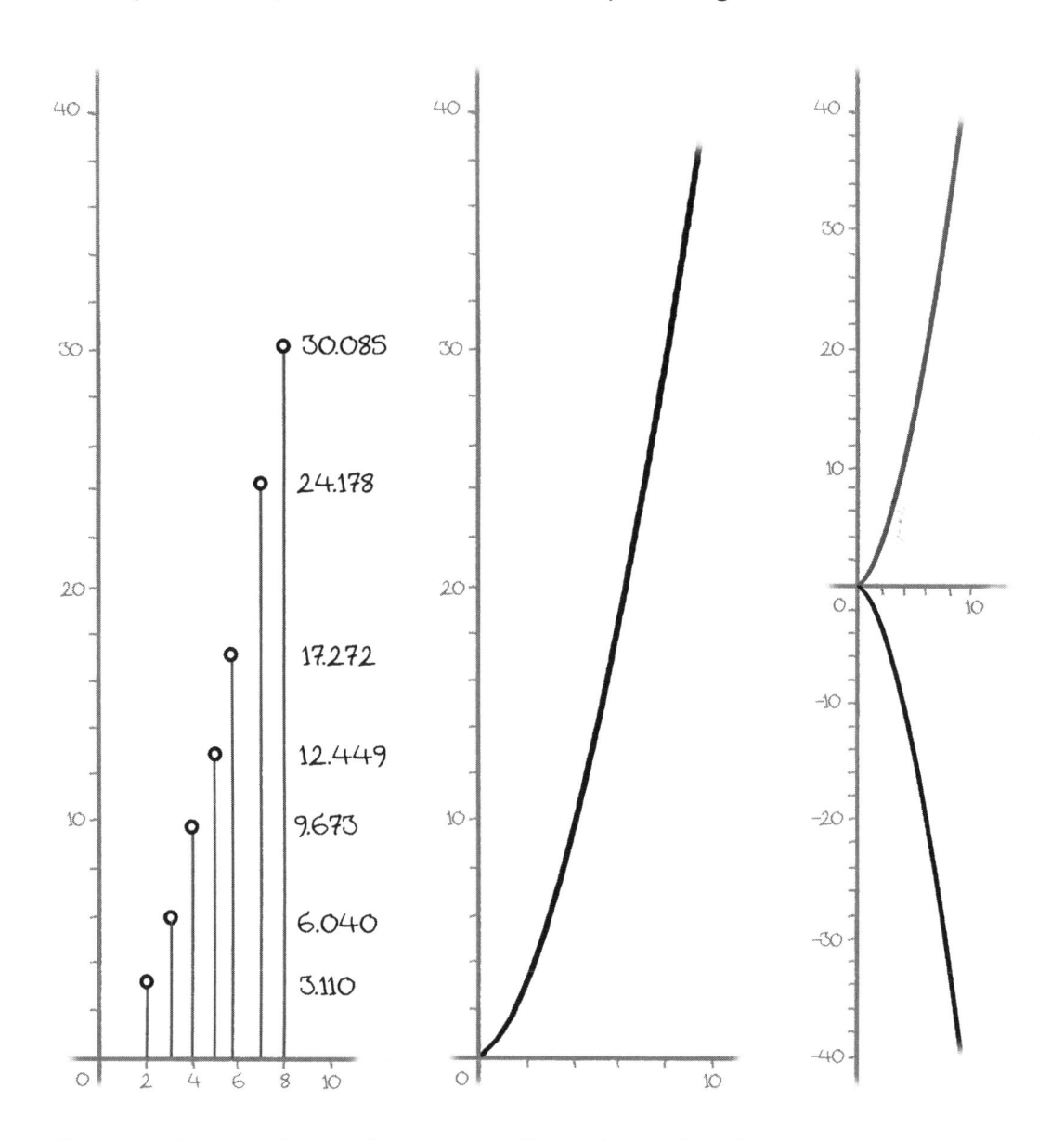

30.085
24.178
17.272
12.449
9.673
6.040
3.110

The second panel shows what you get if you do this for all positive real numbers. The third panel shows a simple reflection of this across the number line. We'll call that the "1.637 power curve"[1]. You could similarly make a power curve for any other number. Here's what some of them look like:

153

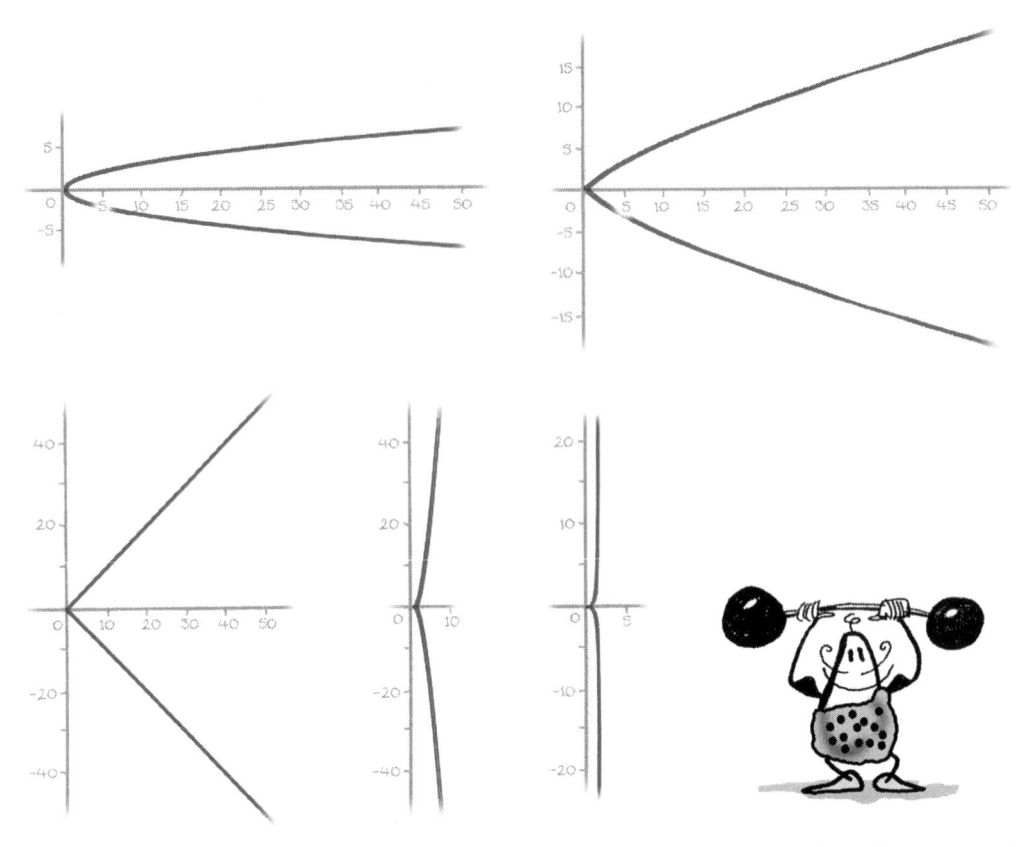

Power curves for powers 0.5 and 0.75 (top row) and for powers 1, 2 and 10 (bottom row).

We'll say that a power curve *encloses* a graph if the graph stays "inside" it all the way along the number line (this is something you can never actually *see* in a graph, but something which you can often prove mathematically about a function).

One very important way mathematicians use to describe how fast a function grows is to *find the power curve with the smallest possible power which, after being "vertically stretched" by some magnitude (if necessary), will enclose the graph of the function.*

I'm using the somewhat vague term "vertically stretched", whereas mathematicians would talk about the power curve being "rescaled". But you should be able to see what's going on – for each point on the curve, we multiply its distance from the horizontal axis by some fixed positive number, pushing the whole thing wider open (or squashing it if our chosen number is less than 1).

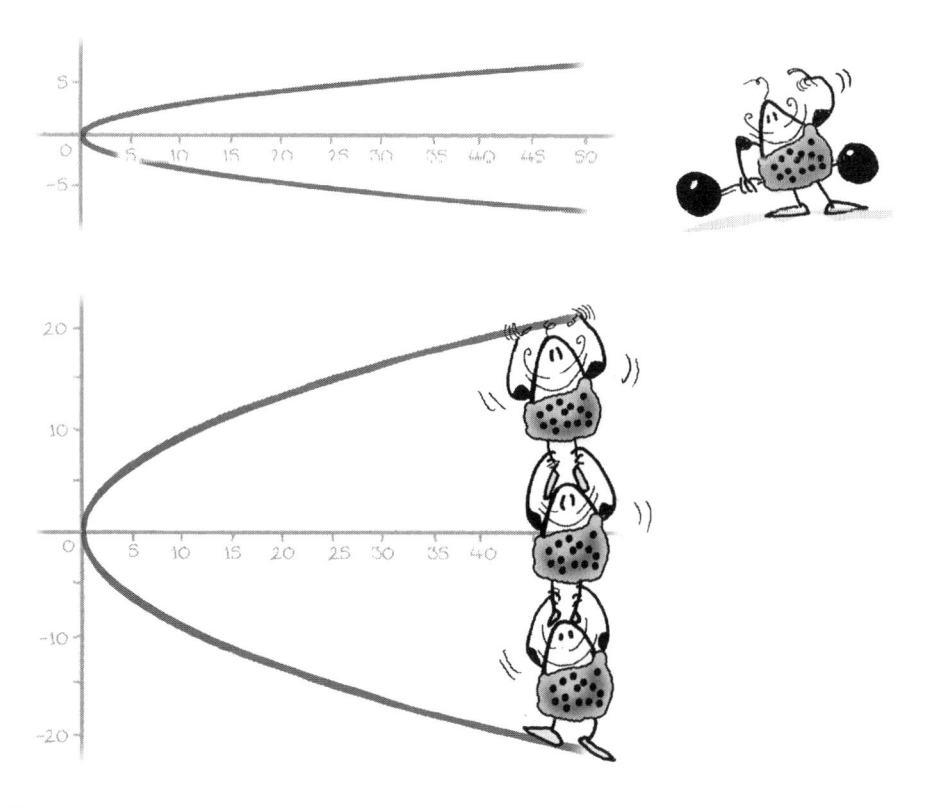

Notice how, for example, the height of the point above 25 is no longer 5, but 15 — here we've stretched the whole graph vertically by a magnitude of 3. We can imagine similarly stretching the graph by various other magnitudes:

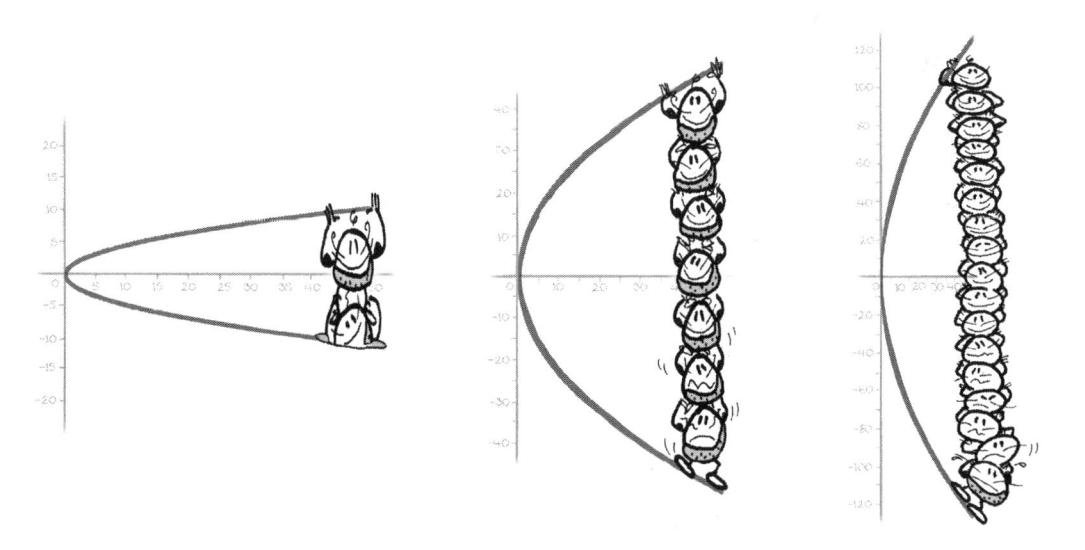

Here we see rescalings by magnitudes 1.5, 7 and 17. Note the heights of the various points above 25.

We now return to the "primeness count deviation" function and its graph. Remember, Bernhard Riemann showed that this severely jagged graph (made of infinitely many diagonal "teeth") can actually be built by adding together an infinite number of perfectly smooth spiral waves, along with a couple of other smooth bits. As a consequence, the amplitude growth rates of these spiral waves are what will determine the rate of growth of the deviation graph.

The RH proposes that all of the nontrivial zeros lie on the critical line, which means that all of the relevant spiral waves will have amplitude growth rate ½. Because of this, it can be shown to be mathematically equivalent to the following statement:

The power curve of any power greater than ½, after some "vertical stretch", will enclose the primeness count deviation graph.

This means that a power of 0.500003, 0.50000000000000017, or whatever, will suffice. How much bigger than ½ doesn't matter, as long as it's a positive amount.

So, if you can prove *that*, then you've proved the RH. And, likewise, if you can *dis*prove it, then you've *dis*proved the RH. To disprove it would involve finding a power bigger than ½ such that its power curve *won't* enclose the graph, regardless of how much you vertically stretch it. This would correspond to finding nontrivial zeta zeros off the critical line – the faster-growing-than-usual spiral waves associated with these would be responsible for the problematic growth of the deviation graph.

Mathematicians would express this reformulation of the RH like this...

$$|\psi(x)-x| = O(x^{\frac{1}{2}+\varepsilon}), \forall \varepsilon > 0$$

...but don't let that trouble you.

The second reformulation of the RH which we'll look at also involves this kind of "bounding" of functions. In particular it involves the *Mertens function*, another "staircase" function whose graph can be built according to the following procedure.

Walk along the number line to 2, the first prime number. Now look at the prime factorization of 2.

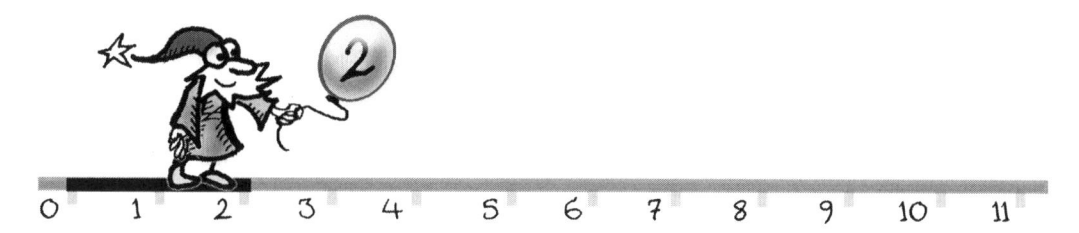

Being a prime number, 2 is "its own factorisation". It has no repeated factors. So it has just **one** factor. 1 is an *odd* number, so go down a step.

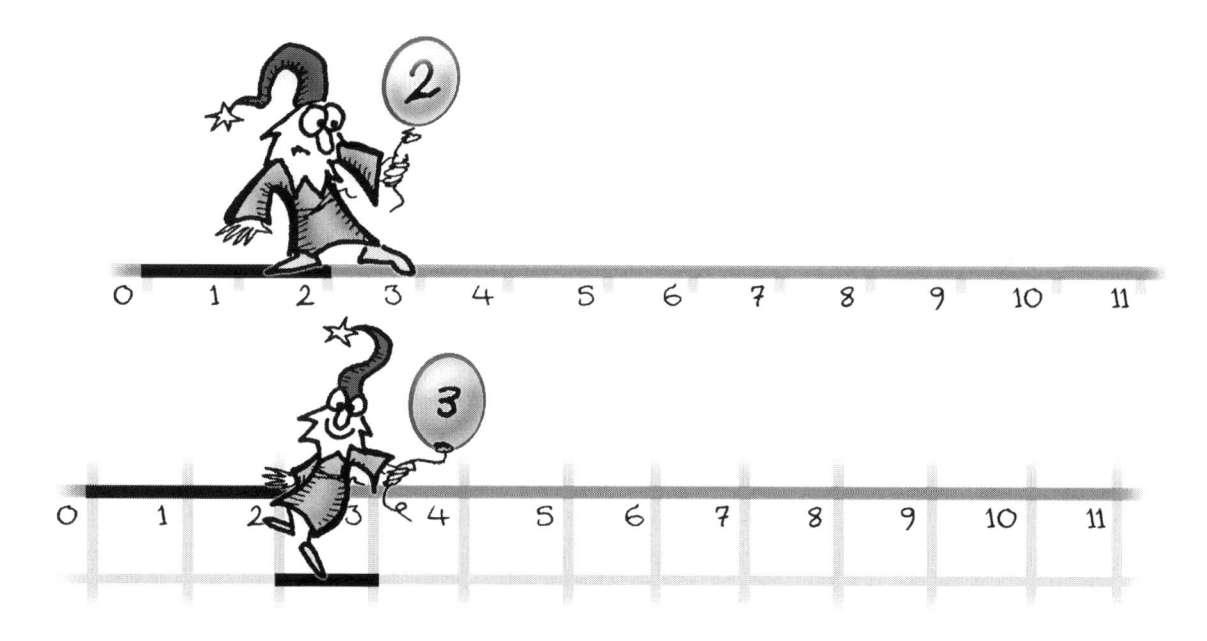

Similarly, 3 is a prime number, so we have just **one** factor, no repeats. 1 is odd, so again go down a step.

157

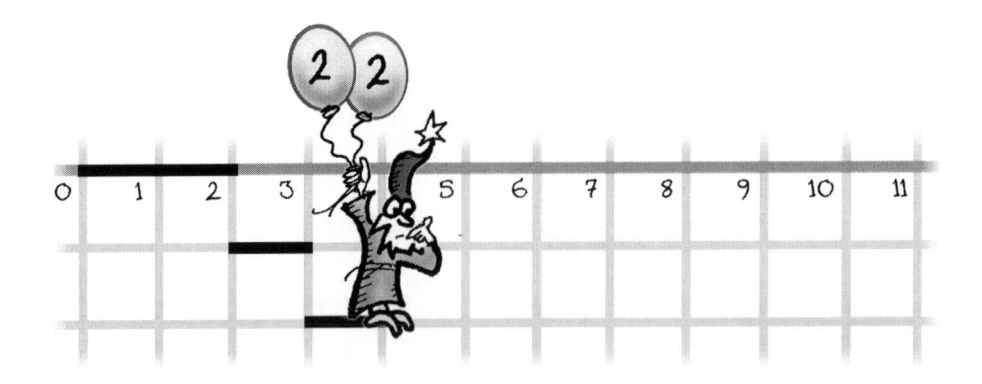

4 is 2 × 2, so we have a repeat factor. Walk on one step. 5 is prime, so go down a step.

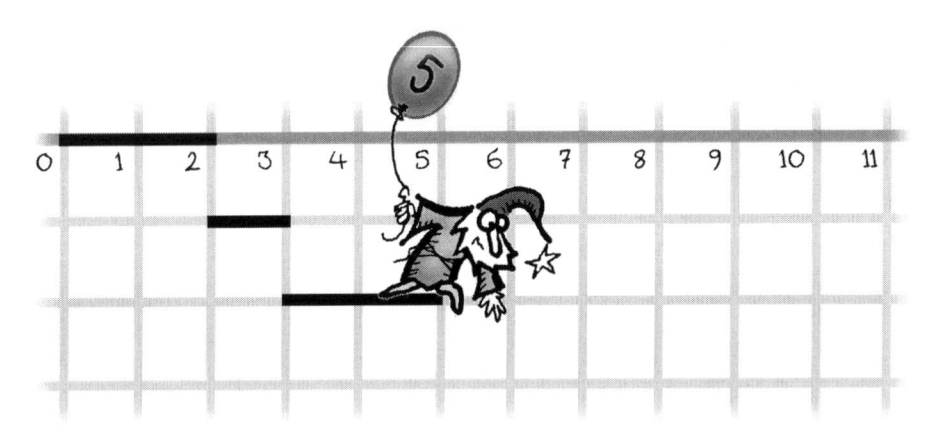

6 is 2 × 3. No repeats. **Two** factors this time. 2 is an *even* number, so this time we go *up* a step.

158

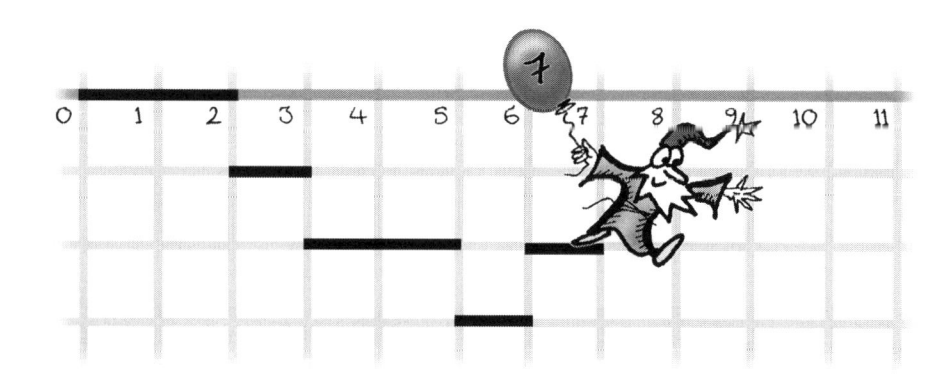

7 is prime, so down a step. 8 is 2 × 2 × 2: we have repeated factors, so walk on.

9 is 3 × 3 – repeated factors. Walk on.

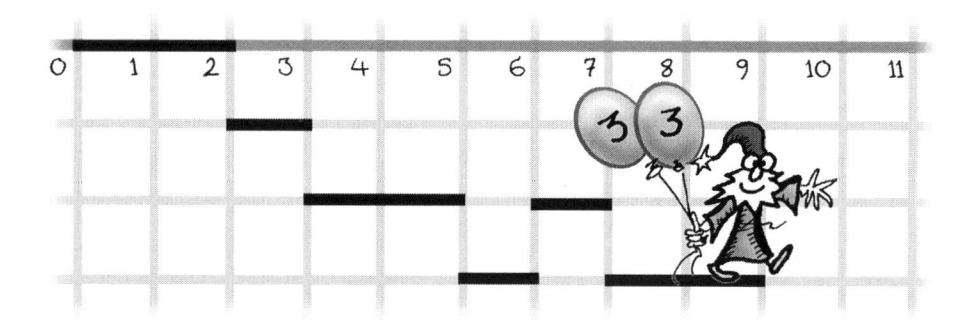

159

10 is 2 × 5, two factors, no repeats, so up a step.

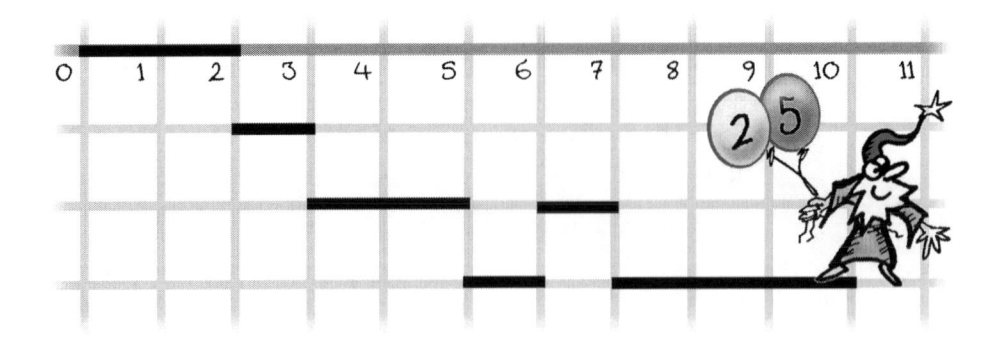

11 is a prime, so down a step.

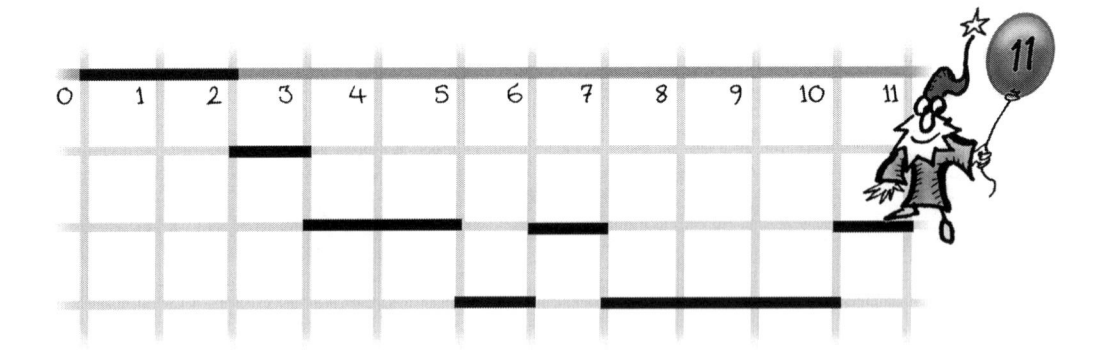

12 = 2 × 2 × 3 − walk on, 13 − down, 14 = 2 × 7 − up, 15 = 3 × 5 − up, 16 = 2 × 2 × 2 × 2 − walk on, 17 − down, 18 = 2 × 3 × 3 − walk on, 19 − down, 20 = 2 × 2 × 5 − walk on, 21 = 3 × 7 − up, 22 = 2 × 11 − up, 24 = 2 × 2 × 2 × 3 − walk on, 25 = 5 × 5 − walk on, 26 = 2 × 13 − up, 27 = 3 × 3 × 3 − walk on, 28 = 2 × 2 × 7 − walk on, 29 − down, 30 = 2 × 3 × 5 − down (there's an odd number of factors), 31 − down, 32 = 2 × 2 × 2 × 2 × 2 − walk on, 33 = 3 × 11 − up… and so on.

You should have the idea by now. At each counting number, we factorise. If there are repeated factors, we walk on. If not, we go either up or down a step, depending on whether the number of factors is even or odd.

This produces the graph of the Mertens function, which has a similar "feel" to the prime count deviation graph:

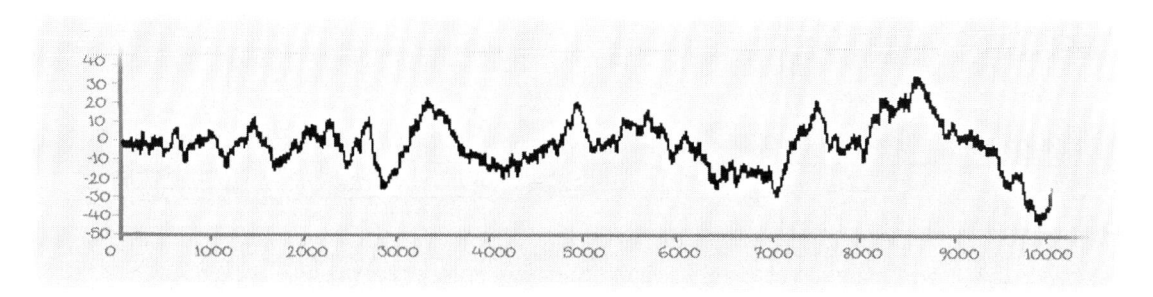

Whereas the prime count deviation graph is made of little pieces of diagonal line, this graph is made of little pieces of horizontal line (although, of course, these aren't visible at this scale).

It turns out that the RH is also equivalent to the following statement:

The power curve of every power greater than ½, after some vertical stretch, will enclose the graph of the Mertens function.

Notice how the graph is displaying a measure of the balance or imbalance between "even" and "odd" factorisations. If there were an overall tendency towards more odd than even factorisations (or *vice versa*), then there would be a general tendency for more "up" steps to occur than "down" steps. The graph then wouldn't be "containable" in the required sense, the statement wouldn't be true, and so the RH would be false. So, interestingly, we see that the RH is equivalent to a statement about a delicate balance in the number system – the balance between odd and even factorisations.

The first reformulation we saw has a similar quality in that it relates the RH to a kind of "balance", that is, no one spiral wave overwhelming the others with its amplitude growth rate. For the RH to be true, they must all have the same rate of growth, the result of all the pairs of nontrivial zeros being "balanced" on the critical line.

The third reformulation also involves the enclosing of a graph by power curves, and it also has this quality of "balance" about it.

If you think back to how the system of rational numbers ℚ was introduced, you might recall that it involved the systematic subdivision of the piece of number line between 0 and 1 into 2, 3, 4, 5, 6, 7,... pieces, whose endpoints gradually fill up the space. So we have this sequence of subdivisions:

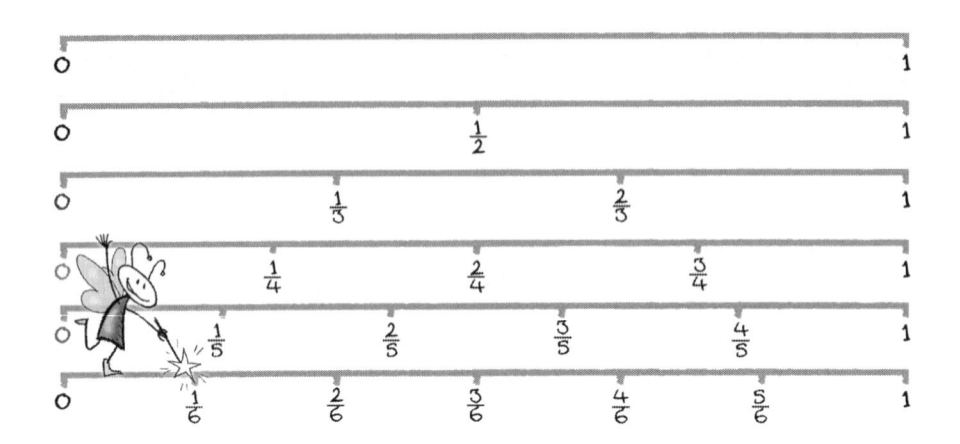

If we show the "running total" of points at each step, we see this:

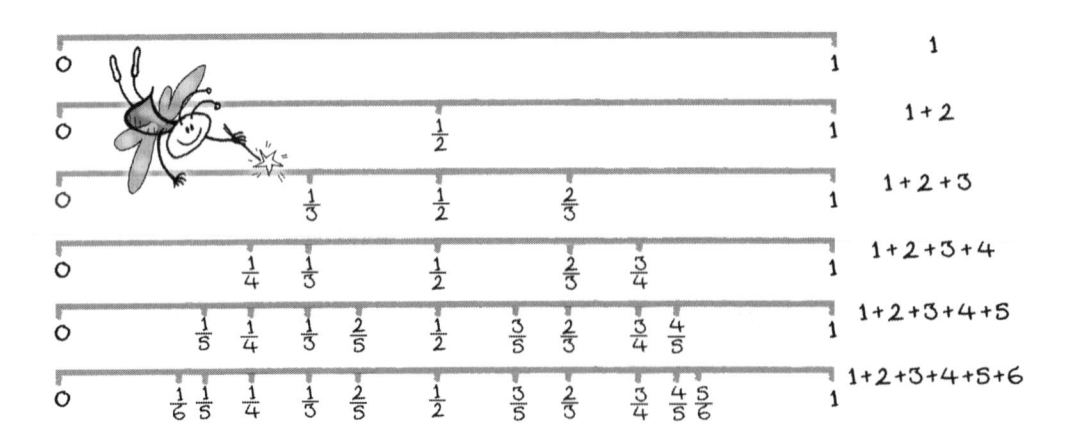

These strings of fractions are called *Farey series*[2]. We can see how they'll gradually fill up the space between 0 and 1, but *we can also see an irregularity*. The points at each stage of the process are *not equally spaced*. For example, after step 6, we have eleven points between 0 and 1:

[Number line showing 0, 1/6, 1/5, 1/4, 1/3, 2/5, 1/2, 3/5, 2/3, 3/4, 4/5, 5/6, 1]

If they were equally spaced, they'd look like this:

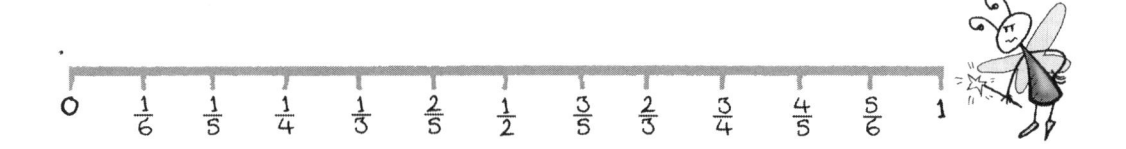

[Number line with the same fractions equally spaced: 0, 1/6, 1/5, 1/4, 1/3, 2/5, 1/2, 3/5, 2/3, 3/4, 4/5, 5/6, 1]

So let's try to come up with a way of measuring "how far from being equally spaced" a list of fractions like this is. The idea is that if they *were* equally spaced, then the answer would be 0 (that's how far they'd be from being equally spaced — "no distance at all"). If we look at the points in order between 0 and 1, we can match them up with their equally-spaced cousins (each one being matched with a point showing where it *would be* if it and its irregular siblings were equally spaced).

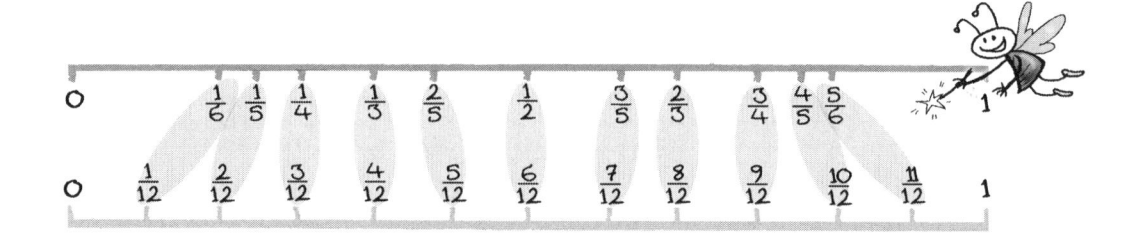

[Two number lines with matching between original fractions (1/6, 1/5, 1/4, 1/3, 2/5, 1/2, 3/5, 2/3, 3/4, 4/5, 5/6) and equally-spaced twelfths (1/12, 2/12, 3/12, 4/12, 5/12, 6/12, 7/12, 8/12, 9/12, 10/12, 11/12)]

We could then measure the distance of each point from "where it should be" according to this scheme. Adding all of these little distances together, we'll end up with something which should work quite well as a measure of "how far from being equally spaced" the fractions are. This method will (as required) give an answer of zero if the points are equally spaced, and the closer they are to being equally spaced (in a vague, intuitive sense), the closer the answer will be to zero.

Let's build a graph from this, the graph of another "staircase" function.

At 1 on the number line, we think of step 1. There are no points to consider here, so the "distance from being equally spaced" is 0. Walk on one step.

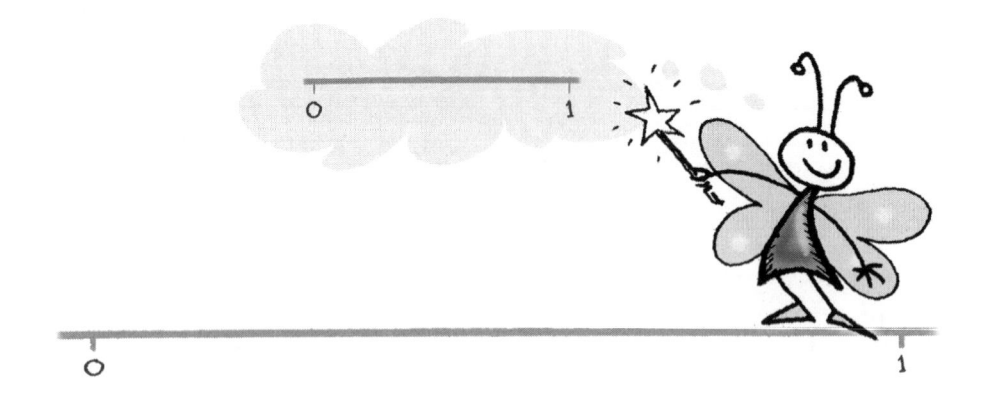

At 2, we think of step 2.

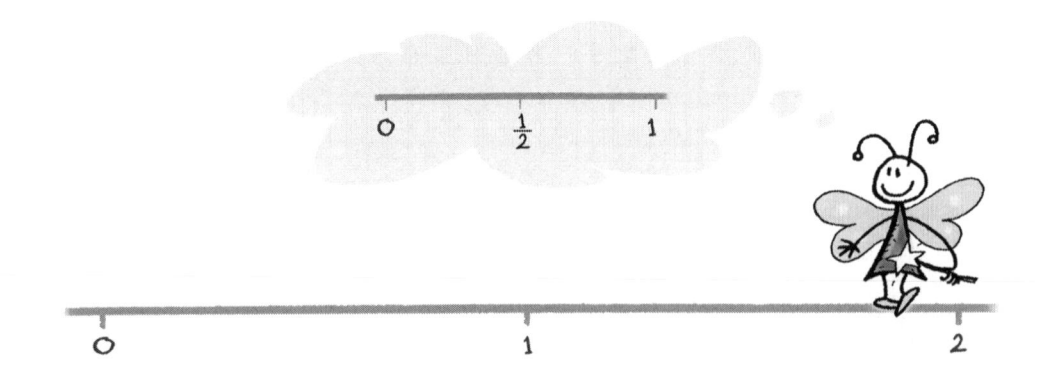

The point in the middle is where it "should" be. We therefore have perfect "equal spacedness", so the "distance" is still 0. Walk on one step.

At 3, we think of the third step. The two new points are *not* where they'd be if they were equally spaced. They're at a distance of 1/12 each from those ideal positions – the total distance is 2/12 or 1/6. So build up your graph at 3 to a height of 1/6:

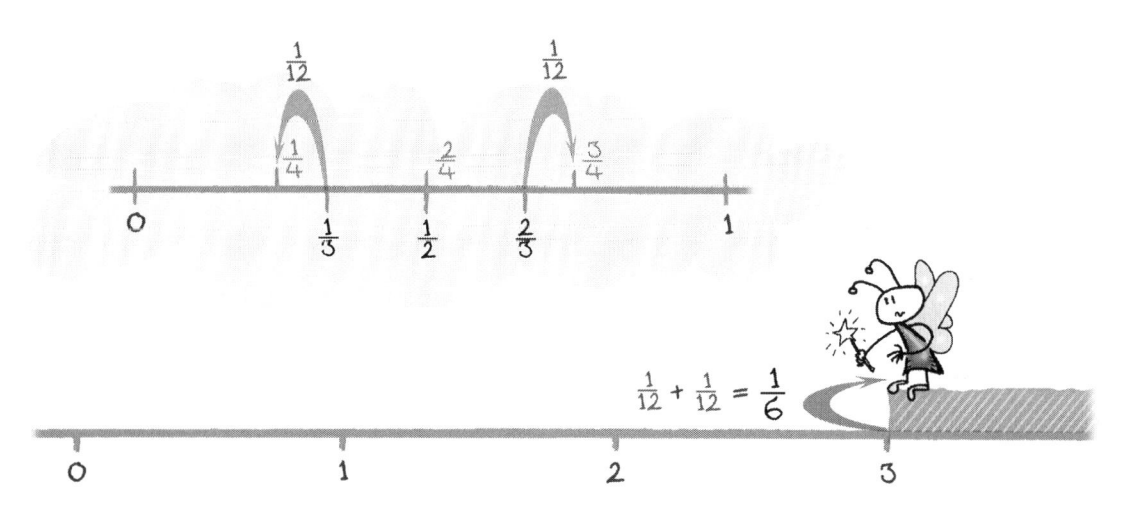

At 4, we have distances 1/12, 0, 0, 0 and 1/12. So the total distance is again 2/12 or 1/6. Build up your graph at 4 to this height (*not illustrated here*).

At 5, the situation is this:

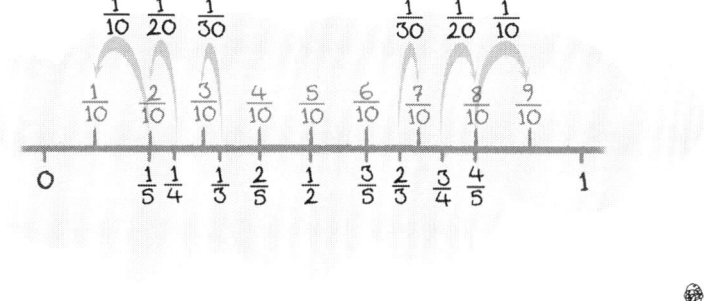

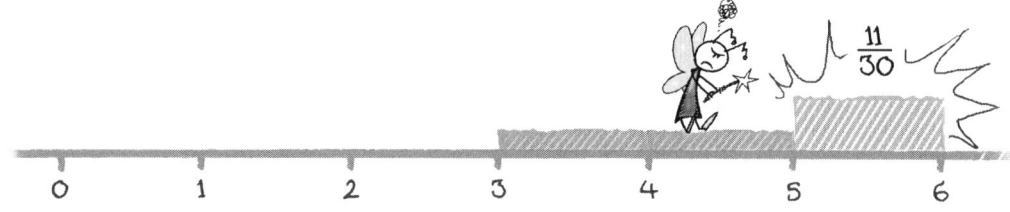

We have distances 1/10, 1/20, 1/30, 0, 0, 0, 1/30, 1/20 and 1/10. This time the total is 11/30. So we build up the graph by this much. And so on.

You might think that this graph wouldn't really grow that much because it's the result of adding increasingly tiny quantities together – but the *amount* of these keeps increasing, so it will. Eventually, we'll end up with something like this:

165

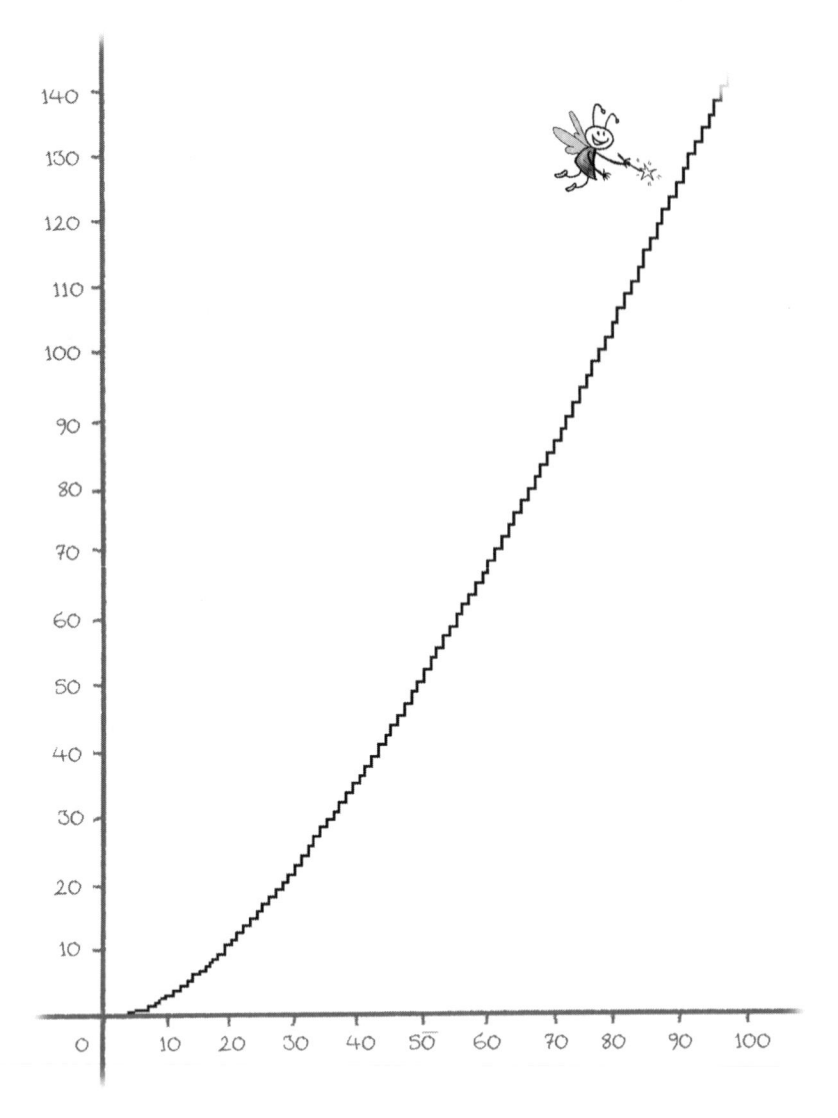

In the 1920s, it was discovered that the Riemann Hypothesis can be reformulated as the following statement:

Every power curve with power greater than ½, after a vertical stretch of some magnitude, will enclose the "Farey" staircase graph just described.

If we continue the process seen on the previous pages, we'll eventually add *any* rational number to our collection – each one must be included at some stage. The rate of growth of the graph shown opposite tells us about how "unevenly" the space between 0 and 1 is filling up with rational numbers. The lower the graph at any stage, the more "evenly" this is occurring. Mathematicians would talk in terms of the "uniform density" of the rational numbers. The RH is effectively the same as saying that "*the rational numbers are distributed as uniformly as possible*" on the number line. The truth or falsehood of the RH can be linked to the truth or falsehood of (a precise version of) this assertion.

So these three examples of RH reformulations all have something to do with "balance" in the number system (in the third one, it's "balance in the distribution of rational numbers on the number line", which involves $\mathbb{Q}$, but then $\mathbb{Q}$ is built directly from the counting numbers).

As I conceded earlier, most RH reformulations couldn't be effectively described in a book like this. They don't always involve "power curves" and rates of growth of functions – it's just that these three all happen to.

In his 2003 popularisation *The Music of the Primes* (and elsewhere), Marcus du Sautoy has informally reformulated the RH in terms of *probability*. To overcome the problem of explaining to his audience what I've called "power curves", and the containment of graphs within vertical rescalings of these, he makes use of a fact concerning the "fairness" or "lack of bias" of a tossed coin. One common test of this, it turns out, involves observing the rate of growth of the deviation from a 50-50 heads/tails split. According to this test, the coin is fair if the deviation graph stays within the "power ½" curve (du Sautoy explains this in terms of the error being less than the square root of the number of tosses). Via this, he's able to relate the RH to the familiar experience of dice throws and coin tosses. He describes a hypothetical random process using the (perhaps somewhat confusing) image of a die with not six sides, but whose number of sides is the *logarithm of the number of throws* (so at the 5000th throw, the probability of the die landing on the side indicating that a prime has been found is not the usual 1-in-6, but now it's 1-in-the-logarithm-of-5000 (that is, 1-in-8.517...).

The Prime Number Theorem can be quite accurately recast in these probabilistic terms, and the Riemann Hypothesis can then be loosely reformulated as proposing that *this hypothetical die is unbiased*. In other words, within the constraints of arithmetic, the prime numbers are as "as randomly distributed as possible".

Yet again, we could say that there's an involvement of a kind of "balance" here. If there is a bias in the process and the primes *aren't* "as randomly distributed as possible", then there must be some sort of emphasis somewhere at the expense of a de-emphasis somewhere else – some sort of "imbalance". Perfect randomness means (very roughly speaking) "nothing is any more likely than anything else", which is a kind of balanced state between all possibilities.

Chapter 25
The Significance of the Riemann Hypothesis

Riemann never used the words "Riemann Hypothesis" – there's an unspoken rule among mathematicians that you never name anything after yourself. You simply publish an account of the (unnamed) theorem, function, hypothesis, conjecture or whatever. If, eventually, it's considered sufficiently important, someone will attach your name to it. So if I were ever to discover an important special function (not something I've ever done), for me to describe it as the "Watkins function" would be considered not just unprofessional, but extremely arrogant. Eventually, though, rather than repeatedly referring to "the function introduced by Watkins in 2011", someone would introduce the shorthand "the Watkins function"[1].

I'm not aware of who first referred to the "Riemann Hypothesis", but I'd guess that this name for it came into use in the early 20th Century, some years before the depth of its significance had been realised. These two words have since taken on a powerful resonance which even Riemann probably couldn't have anticipated.

THE WIDER MATHEMATICAL CONTEXT

That the Riemann Hypothesis is related to so many diverse areas of mathematics can also be seen in the extent to which new mathematics has been "built on top of it". There are many, many mathematical statements which aren't *equivalent* to the RH, but *are* dependent on it.

The idea is that if the RH is true, then these mathematical statements will also be true, but *not vice versa*. That is, it's possible for them to be true *without* the RH necessarily having to be true. So, these statements are often presented as theorems which begin "Assume the Riemann Hypothesis" (that is, assume that it's true). You can prove them to be "true if the RH is true". Therefore, these theorems are currently known to be "provisionally true" – their truth would follow from the truth of the RH. If, eventually, someone is able to prove the RH, then the status of all of these theorems will instantly change from "provisionally true" to *absolutely* true. This is another reason why mathematicians would like to see the RH proved. The situation would be something like a lot of people chipping away at a cliff face for decades, and then a landslide suddenly occurring. Proof of the RH would cause this sort of "mathematical landslide".

Earlier, I hinted that the Riemann zeta function is just one member of a whole family of "zeta functions". It can also be generalised in a way which produces another family of what are called "*L*-functions"[2]. These various zeta and *L*-functions generally have certain features in common with Riemann's zeta function. They're all functions defined on $\mathbb{C}$ (minus a handful of isolated points, as with the pole of the Riemann zeta function at 1). Their zeros tend to be their feature of greatest interest. Often these functions have a "functional equation" along similar lines to Riemann's (see page 72). In many cases, the zeros appear to lie on a vertical line (or, in some cases, on a circle), but this is generally an unproved observation (as the tendency for the zeros of the Riemann zeta function to lie on the critical line is an unproved observation). For this reason, many zeta and all *L*-functions have an associated "Riemann Hypothesis". In a couple of cases, these "Riemann Hypotheses" have even been proved[3], which can be taken as a kind of indirect evidence for the truth of the "classical" RH (the one we've been looking at).

There are also things known as the "generalised RH" and the "Grand RH" which state that the RH's of a whole range of zeta and *L*-functions are *all* true. To prove either of these would not just prove the classical RH, but would go a lot further. And the mathematical "landslide" of theorems then known to be true would be considerably larger.

> *"Over the years striking analogies have been observed between the Riemann zeta-function and other zeta- or L-functions. While these functions are seemingly independent of each other, there is growing evidence that they are all somehow connected in a way that we do not fully understand. In any event, trying to understand, or at least classify, all of the objects which we believe satisfy the Riemann hypothesis is a reasonable thing to do."* (J. Brian Conrey)[4]

Conrey's statement is necessarily imprecise. This kind of writing is not formal mathematical reasoning but rather a kind of informal descriptive account of the issues involved. It's the sort of thing found in the introductory sections of formal academic papers, as well as in articles written for the benefit of nonspecialists. He's using the imprecise word "objects" rather than specifically "functions", presumably to leave open the possibility that something like a "Riemann Hypothesis" could be defined for mathematical entities more general than functions.

Rather than the details, the thing worth taking from this is the strangeness of the situation. First, I showed you a function, the Riemann zeta function, presenting it as something utterly central to the workings of the number system, a unique, "monolithic" mathematical entity. It's just *there*, underlying the positive integers. I then went on to tell you that it's *just one member of a family*. So, if Riemann's zeta function underpins the number system, what are all these other family members doing? To borrow Brian Conrey's words, this is something "we do not fully understand". In the words of a few other mathematicians:

> *"Whenever entities are counted with some mathematical structure on them, it is likely that a zeta function can be set up... Zeta functions show up in all areas of mathematics and they encode properties of the counted objects which are well hidden and hard to come by otherwise. They easily give fuel for bold new conjectures and thus drive on mathematical research. It is a fairly safe assertion to say that zeta functions of various kinds will stay in the focus of mathematical attention for times to come."* (Anton Deitmar)[5]

"...it seems that [these zeta functions] provide some of the best reasons for believing that the Riemann hypothesis is true – for believing, in other words, that there is a profound and as yet uncomprehended number-theoretic phenomenon, one facet of which is that the [Riemann zeta zeros] all lie on [the critical line]." (Harold M. Edwards)[6]

"Some decades ago I made – somewhat in jest – the suggestion that one should get accepted a non-proliferation treaty of zeta functions. There was becoming such an overwhelming variety of these objects." (Atle Selberg, quoted by K. Sabbagh 2003)[7]

In recent years, mathematicians like Andreas Juhl have been carrying out research aimed at showing that these functions are all manifestations of a single principle[8]. I think it would be fair to say that Riemann's is widely perceived as the "root" or "mother" of all of the zeta and *L*-functions – the "archetypal" zeta function.

MUSICAL CONNECTIONS

Quite a few musical metaphors have been inspired by the harmonic structure which Riemann revealed to underlie the number system (we've seen a couple already). Centuries ago, European cosmologists spoke of "the music of the spheres", based on the idea that the solar system was a perfect mathematical creation, a manifestation of divine will and cosmic harmony. The planets in their motions were imagined to produce (inaudible) harmonic sounds. This notion reflected certain ideas held at the time about the nature of the Universe. As astronomical knowledge has increased and we've come to terms with the seemingly arbitrary, "imperfect" nature of the solar system, less and less has been said about "the music of the spheres". It's now part of the poet's vocabulary, where once it was a scientific or philosophical notion. But we now have people referring to the "music of the primes" in much the same way. Typifying this, Marcus du Sautoy called his 2003 book on the RH *The Music of the Primes*.

This is interesting. We're still seeking manifestations of cosmic harmony and divine will in our world, but now we're not looking outwardly into space. Instead we're looking inwardly into the number system. This switch is perhaps evident in the following observation:

> "*I sometimes have the feeling that the number system is comparable with the universe that the astronomer is studying... The number system is something like a cosmos.*"
> (M. Jutila)[9]

The aptness of the "large telescopes" metaphor which I used when discussing the application of powerful computers to the search for prime numbers (back in Chapter 7) also follows this analogy.

It seems to me that there's a hidden or unconscious belief evident here that some sort of "creator" is responsible for something we see before us, and that as this creator is in some sense perfect, we should expect to find perfection or harmony in the thing we're looking at (once the solar system, now the number system).

As we've seen, Enrico Bombieri has written of an "*arcane music and secret harmony composed by the prime numbers*", as well as introducing the metaphor of an instrument drowning out the rest of an orchestra.

Michael Berry and Jon Keating, two mathematical physicists based at Bristol University (who we'll hear a lot more about in Volume 3), used the sequence of prime numbers to construct a very particular function which, if the RH is true, has a "discrete spectrum" (that is, it breaks down cleanly into a sequence of sine waves) and, if the RH is false, doesn't. They make the following remark:

> "*The frequencies...are reminiscent of the decomposition of a musical sound into its constituent harmonics. Therefore there is a sense in which we can give a one-line non-technical statement of the Riemann hypothesis: "The primes have music in them.*""[10]

This might seem slightly confusing since we know that the distribution of primes – more correctly, the deviation of the distribution of primes from its own "average behaviour" – can still be decomposed in terms of a set of "harmonics" even if the RH fails. However, as Bombieri has suggested, if the RH is false, then such a harmonic decomposition would be "aesthetically distasteful" (and so, arguably, the primes wouldn't "have music in them").

Berry and Keating have constructed a somewhat artificial context in which the question of the RH can be reduced to "music versus no music". The slightly convoluted nature of the function they constructed suggests that there was perhaps, consciously or otherwise, a wish to introduce a musical analogy into their narrative.

In any case, Michael Berry was sufficiently curious to generate the sound in question, as he reported to popular science writer Erica Klarreich:

> "I've tried to play this music by putting a few thousand primes into my computer...but it's just a horrible cacophony. You'd actually need billions or trillions – someone with a more powerful machine should do it."[11]

Perhaps someone has, but I'm not quite sure why billions or trillions of primes would make the sound any less cacophonic, as that would still only represent an infinitesimal proportion of them (this is unavoidable, however many we compute).

In a 1998 article aimed at a general readership, Marcus du Sautoy (an enthusiastic musician as well as an accomplished mathematician) wrote this:

> "...mathematicians like to look for patterns, and the primes probably offer the ultimate challenge. When you look at a list of them stretching off to infinity, they look chaotic, like weeds growing through an expanse of grass representing all numbers. For centuries mathematicians have striven to find rhyme and reason amongst this jumble. Is there any music that we can hear in this random noise?"[12]

He took this metaphor further, writing a much more extensive popular account of the RH in the form of his critically acclaimed *The Music of the Primes*, full of references to music such as these:

> "Gauss had heard the first big theme in the music of the primes. But it was one of his few students, Riemann, who would truly unleash the full force of the hidden harmonies that lay behind the cacophony of the primes."[13]

> "For centuries, mathematicians had been listening to the primes and hearing only disorganised noise. These numbers were like random notes wildly dotted on a mathematical stave, with no discernible tune. Now Riemann had found new ears with which to listen to these mysterious tones. The sine-like waves that Riemann had created from the zeros in his zeta landscape revealed some hidden harmonic structure."[14]

The "sine-like waves" are of course the "spiral waves" we've been considering.

> "*Until* [the RH is proved], *we shall listen enthralled by this unpredictable mathematical music, unable to master its twists and turns. The primes have been a constant companion in our exploration of the mathematical world yet they remain the most enigmatic of all numbers. Despite the best efforts of the greatest mathematical minds to explain the modulation and transformation of this mystical music, the primes remain an unanswered riddle.*"[15]

The recurring popularity of the music analogy perhaps reveals something about what we expect, hope or want from the number system.

I expect that various schools of psychological theory would have interesting interpretations to offer about this matter. It has something to do with a "collective psychological" or "cultural" relationship with the number system. Humans have a tendency to associate music with the process of creation, from the Aboriginal Australian idea of "singing the world into being" to European Christians in various centuries picturing their Creator God surrounded by angels singing and playing harps. Perhaps we're touching on some sort of "creation myth" underlying the number system (which underlies the world in some way, or at least the world *as we currently experience it*).

A more mundane analysis would argue that the involvement of sine-like waves naturally leads some people to think of music (which is, in a sense, built out of sine waves). The universal and nonphysical nature of the number system then leads them to set this music apart from Earthly music, so we have a link with "heavenly music" = "music of the heavens" = "music produced by the geometry of the solar system" = "music of the spheres".

It's worth remembering that in the earliest days of university education, the syllabus was *universal* (unlike the ultra-specialisms of today) and consisted of what was called the *quadrivium*: arithmetic, geometry, music and astronomy. At that time, God was being conceptualised as some kind of supreme "architect" or "geometer". These connections between number, time, space and vibration will be explored further in the final volume.

All of this helps to explain my motivation for introducing the quasi-religious imagery in Volume 1, with the trumpeting angels bringing the distribution of primes into being.

Another, less traditionally "reverent" image we might introduce would be...

...an infinite-fingered Goddess bashing out an infinitely crazy jazz chord containing infinitely many notes on an infinitely long "critical strip" piano keyboard. Out of this chaotic sound emerges the sequence of primes, the skeleton of the number system, supporting the rest of the positive integers from which all of mathematics ultimately follows.

HISTORICAL OVERVIEW

What we could call "the mystery of the primes" is largely captured in the way that the Riemann Hypothesis seems to obstinately resist being proved. And this isn't from any lack of effort. Some of the most mathematically talented minds ever to exist have been grappling continually with the problem since it was posed.

But even if someone were to prove it tomorrow, that wouldn't be the end of the matter. A belief is often expressed that the proof of the RH may open up another, deeper layer of mystery:

> "Proving the Riemann hypothesis won't end the story. It will prompt a sequence of even harder, more penetrating questions. Why do the primes achieve such a delicate balance between randomness and order? ...what other jewels will we uncover when we dig deeper? Those who believe mathematics holds the key to the Universe might do well to ponder a question that goes back to the ancients: What secrets are locked within the primes?" (E. Klarreich)[16]

This impasse we've reached with the RH is just "where we're at" with the number system *at this point in history* (2011, when I'm writing this). It's worth taking a step back from a time "line" and assessing where we are:

Certain landmarks are visible, but humanity is very much in the fog at the moment when it comes to these issues. However much we think we know now, it's quite possible that we'll seem naive and ignorant from the perspective of the distant future.

So what's the status of the Riemann Hypothesis in 2011?

A century ago, in 1911, just over fifty years had passed since Riemann had modestly proposed his Hypothesis, and its true significance was only just being appreciated by the global community of mathematicians. This would be unimaginable now, with mathematical ideas circulating via e-mail, "preprint" articles being published online and mathematicians frequently using air travel to attend international conferences. At that time, however, international mathematics communication occurred at the speed of the postal system, so the awareness of new ideas tended to be more regionalised.

In 1907, John Littlewood was a postgraduate student at Trinity College, Cambridge, being supervised by Ernest Barnes. Barnes suggested the Riemann Hypothesis as a problem for Littlewood to solve (to do this with a postgraduate in 2011 would be cruel and absurd). Obviously, Littlewood never proved or disproved the RH, but he did go on to make some significant contributions to the theory of the Riemann zeta function.

In 1909, Edmund Landau published a classic book on analytic number theory[17] through which the mathematical world fully awoke to the realities of the mysterious zeta function lurking behind the number system and the enigma that is the Riemann Hypothesis. Littlewood continued to work on the problem, as did his Trinity College colleague Godfrey Hardy, and, both together and individually, they went on to publish numerous papers on the zeta function. Still, the mathematical community was not brought substantially closer to a proof as a result of their efforts. In 1962, an essay by Littlewood was published in which he explained why he believed the RH to be false[18]. This would almost certainly have been a minority opinion at that time, and even more so in recent years. This could have been an honest reflection of his mathematical instincts, or possibly a psychological reaction, an expression of his disappointment at being unable to find a proof.

A burst of interest in the problem occurred at the end of the 20th century. Major worldwide conferences on the RH and related matters, sponsored by the American Institute of Mathematics, were held in 1996 (Seattle), 1998 (Vienna) and 2002 (New York).

Until quite recently, the RH was hardly known of outside the mathematical community. In 2001, though, it made a cameo appearance in that year's Academy Award-winning Hollywood film *A Beautiful Mind*, in which Russell Crowe plays an alarmingly fictionalised version of real-life – and *living* – mathematician John Nash, struggling with both mental illness and the Hypothesis. The film contains an extraordinary scene wherein an incoherent Nash delivers a stream-of-consciousness lecture to a bewildered audience, in which he claims that the zeros of the Riemann zeta function correspond to "*singularities in space-time*". Although this reference would have passed most viewers by, it was striking to see the particularly obscure interface of prime number theory and physics (a major theme in Volume 3) surfacing in such a widely viewed film.

Over the next couple of years, three popular books about the RH were published: Marcus du Sautoy's *The Music of the Primes*, John Derbyshire's *Prime Obsession* and Karl Sabbagh's *Dr. Riemann's Zeros*. All three books are worth reading, each with its own style and emphases, and they're the ideal resources if you want to know more about the historical and cultural aspects of the RH. The sales of these books will have been helped along by the fact that in 2000, an American philanthropic organisation, the Clay Mathematics Institute, offered $1000000 prizes for seven notorious unsolved mathematical problems including, of course, the RH. Articles about the "million dollar math puzzle" (and suchlike) have since appeared in the popular media, with varying degrees of descriptive success. It's not easy to convey the essence of the problem in a few paragraphs, as you'll have realised by now!

Although a million dollar prize fleetingly grabs the attention of the masses, both professional and amateur mathematicians would still be desperately pursuing a proof without any financial incentive. As the historian of mathematics Eric Temple Bell wrote about the Hypothesis in 1937, "*Whoever proves or disproves it will cover himself in glory...*" [19]

More recently, Hugh Montgomery, whose contributions to the theory of the zeta function will be discussed in the next volume, said in an interview:

> "So if you could be the Devil and offer a mathematician to sell his soul for the proof of one theorem – what theorem would most mathematicians ask for? I think it would be the Riemann Hypothesis." [20]

The intrinsic fascination with the problem is sufficient motivation for the community of mathematicians and mathematical physicists involved in the quest for a proof of the RH. And it is appropriate to call this a "quest", since the problem has taken on an almost mythical quality, as evidenced in a passage from the cover of Derbyshire's book:

> "Hunting down the solution to the Riemann Hypothesis has become an obsession for many – the veritable 'great white whale' of mathematical research. Yet despite determined efforts by generations of mathematicians, the Riemann Hypothesis defies resolution." [21]

Adding to this tendency was the title of a fourth popular book on the RH which appeared in 2005, Dan Rockmore's *Stalking the Riemann Hypothesis: The Quest to Find the Hidden Law of Prime Numbers* [22]. The following quotations also arguably tie in with this sense of a "quest":

> "...the Riemann hypothesis remains one of the outstanding challenges of mathematics, a prize which has tantalized and eluded some of the most brilliant mathematicians of this century...Hilbert is reputed to have said that the first comment he would make after waking at the end of a thousand year sleep would be, 'Is the Riemann hypothesis established yet?'" (Richard E. Bellman) [23]

> "We still await the person whose name will live for ever as the mathematician who made the primes sing." (Marcus du Sautoy) [24]

It feels to me like something is coming together. I can't back this up with any evidence, but having been following the problem since the late 1990s, I have a sense of some sort of major historical confluence approaching. Contributing to this feeling are the occurrence of the 1996, 1998 and 2002 conferences, the sudden public interest resulting from the popular books published in 2002–3 and the sheer volume of literature which is continually emerging, relating to different aspects of the problem.

Michael Berry, someone considerably more qualified to comment on this than I, has expressed a similar feeling:

> "I have a feeling that the hypothesis will be cracked in the next few years. I see the strands coming together. Someone will soon get the million dollars."[25]

Unlike the much more famous "Fermat's Last Theorem" which was proved by Andrew Wiles in 1993 (the subject of popular books, articles, documentaries and even a Broadway musical), the RH is truly fundamental to the core issues of mathematical reality. Fermat's Last Theorem was, in comparison, a mathematical amusement of marginal value (although the search for its proof admittedly led to a lot of important new mathematics).

Here's what three different mathematicians have had to say in recent years:

> "Hilbert included the problem of proving the Riemann hypothesis in his list of the most important unsolved problems which confronted mathematics in 1900, and the attempt to solve this problem has occupied the best efforts of many of the best mathematicians of the twentieth century. It is now unquestionably the most celebrated problem in mathematics and it continues to attract the attention of the best mathematicians, not only because it has gone unsolved for so long but also because it appears tantalizingly vulnerable and because its solution would probably bring to light new techniques of far-reaching importance." (Harold M. Edwards, 1974)[26]

> "It remains unresolved but, if true, the Riemann Hypothesis will go to the heart of what makes so much of mathematics tick: the prime numbers. These indivisible numbers are the atoms of arithmetic. Every number can be built by multiplying prime numbers together. The primes have fascinated generations of mathematicians and non-mathematicians alike, yet their properties remain deeply mysterious. Whoever proves or disproves the Riemann Hypothesis will discover the key to many of their secrets and this is why it ranks above Fermat as the theorem for whose proof mathematicians would trade their soul with Mephistopheles." (Marcus du Sautoy, 1998)[27]

"I would trade everything I know in mathematics for just knowing the proof of the Riemann Hypothesis. It's just gorgeous stuff. I'm only worried that what may happen is that a proof will be given by somebody and I will be unable to understand it. That's the worst..." (Henryk Iwaniec, 2002)[28]

HOW THINGS NOW STAND

At this point, it would seem reasonable for me to tell you a bit about who is currently considered "in the running" to prove the Riemann Hypothesis. I'm hesitant to do this, partly because I can only tell you about the situation in 2011 – things will move on, and at the time you're reading this, anything I'll have written will be out of date to some greater or lesser extent. Also, what you could call the "human side" of the RH has been extremely well documented (at least up until a few years ago) in the four popular books on the subject (Sabbagh, Derbyshire, du Sautoy and Rockmore).

But as I've been following the situation since the late 1990s, maintaining an extensive website on the subject and various related matters, I feel I should at least give a general overview.

There have been false alarms, April Fools jokes, comedy proofs and "theological" or "philosophical" arguments earnestly presented as "proofs" by non-mathematicians (and consequently ignored by mathematicians). As of 2011, if you were to look on the Internet, you'd find a dozen or more *claimed* proofs of the Riemann Hypothesis. These are generally the work of amateurs and are transparently flawed to those mathematicians working in relevant areas of the subject.

There are also a couple of researchers to have offered proofs of the RH who do work within academia, but whose ideas would be considered "fringe" by the majority of the mathematical community. In both cases, proofs have been made available online, then withdrawn and revised multiple times. They have a few supporters who would be taken more "seriously", researchers who think that at least the beginnings of a successful approach to the problem may have been found. But, for the most part, they struggle to get anyone to look carefully at their work.

One of these two joked to me that it's harder to get someone to read your proof of the RH than it is to actually prove the RH! There's a general disbelief within the mathematical community that anyone outside a tiny handful of mathematicians who are "widely recognised" to be the true experts could come up with a valid proof. There's also an instinctive belief (or prejudice?) that certain approaches couldn't possibly lead to a proof of the RH.

This isn't mathematics we're discussing now, it's more like a subtle web of unspoken "politics" and/or belief within the mathematical community. Intuitions, hunches, speculations, gut feelings and prejudices are just as common among mathematicians as among everyone else, and they ultimately affect the directions in which mathematical research proceeds. But *they're not part of mathematics*. They're part of some sort of "psycho-mathematical interface" which very little has been written about (yet).

There are also a few "disproofs" in circulation (at various levels of professionalism), although none of them involves having found a zeta zero off the critical line. Rather, they involve various logical/mathematical arguments as to why the RH must be false (that is, why all the zeros cannot possibly be on the critical line). Again, these are ignored by the mathematical community, the general feeling being that if they were correct, someone "serious" would have noticed by now and brought the work to widespread attention. Also, as I've already mentioned, the overwhelming opinion is that the RH is going to be true.

If bookmakers were to offer odds on who was going to prove the RH (which, as far as I know, they don't in 2011) there are a few names I could think of which would certainly show up. But the situation is unclear because it's possible to carry out mathematical research entirely in secret, as Andrew Wiles did while working on his proof of Fermat's Last Theorem. Working secretly on a problem eliminates the possibility of someone else taking your partially-formed idea and completing it before you do. This comes from a desire (perfectly natural, some would say) to be *the one*, to get the recognition, glory, respect, profound sense of satisfaction or prize money. So, at any time, there could be someone who isn't even publicly known to be working on the RH but is on the verge of completing a proof.

Michel Lapidus, a French mathematician who has been working at the University of California for some years, recently published a book called *In Search of the Riemann Zeros*[29]. The title, unusual for an academic mathematics text, betrays a sense of mystery which the author clearly feels in connection with this subject (this can be witnessed in his choice of language and quotations throughout the book). His research brings together many different threads, both purely mathematical and physics-inspired, from the century-and-a-half-long international quest to prove the Riemann Hypothesis. But Lapidus isn't claiming to have a proof. Instead, he's putting forward some ideas which may (over who-knows-what time scale) feed into the effort to construct a proof. One of the key ideas he presents is that of a "flow", in particular, a *noncommutative flow on a moduli space of fractal membranes*(!) This is, as you'd expect, highly specialised stuff, the expression of which looks like this...

> More accurately, the modular automorphism flow $\{\sigma_t^\varphi\}_{t \in \mathbb{R}}$ (along with its analytic continuation) on the type III$_1$ factor $\mathcal{M}_{fm}$ associated with this weight φ can be thought of as an appropriate substitute for (and extension of) the notion of Frobenius flow. Similarly, the moduli space of fractal membranes $\mathcal{M}_{fm}$ can itself be viewed as a natural extension of and substitute for Deninger's 'arithmetic site'. (See Remarks 5.4.6(a) and 5.4.8(a), along with §5.4.2e and §5.5.)
>
> (iv) Accordingly, $\mathcal{M}_{fm}$ can be thought of as a suitable noncommutative deformation of the arithmetic site (i.e., of the space of 'all' arithmetic geometries). Finally, as will be further discussed in §5.4.2, §5.5.2 and §5.5.3, the flow of zeta functions (or partition functions) induced by the modular flow on $\mathcal{M}_{fm}$ is asymptotically 'pushing' towards (or against) the space of all arithmetic zeta functions associated with $\mathcal{M}_{fm}$, viewed as the arithmetic site.

...so I won't even begin to explain, but the word "flow" is the key here. The framework that Lapidus has in mind in some sense *sets the Riemann zeros "in motion"*. Their familiar arrangement in the critical strip is understood as corresponding to one fixed, stable location within a "flowing", evolving space, a type of (pure mathematical) "dynamical system". And, because of the way they're related to the zeros, Lapidus' flow could be understood as setting the prime numbers "in motion" too [30].

It's too early to tell, but I have a strong sense that Lapidus has brought something really new and important to the search for a proof of the RH. Even if I'm wrong, it seems that some similarly imaginative new idea is going to be needed. The following (informed) opinions all suggest that the current inability to prove the RH is due to the lack of a necessary new idea:

> "*...number theorists say they are at least one 'big idea' away from even the beginnings of a proof. Mathematicians aren't yet sure what to aim at, says* [Princeton University mathematician Peter] *Sarnak.*" (Barry Cipra) [31]

> "*...the Riemann Hypothesis will be settled without any fundamental changes in our mathematical thoughts, namely, all tools are ready to attack it but just a penetrating idea is missing.*" (Y. Motohashi, quoted by K. Sabbagh, 2003) [32]

> "*...there have probably been very few attempts at proving the Riemann hypothesis, because, simply, no one has ever had any really good idea for how to go about it!*" (Atle Selberg) [33]

> "*I still think that some major new idea is needed here.*" (Enrico Bombieri) [34]

The implication is that if/when this important new idea surfaces, everything will suddenly fall into place: a proof of the RH will be quickly constructed based on the idea, and quite possibly our whole understanding of the number system will be revolutionised in the process.

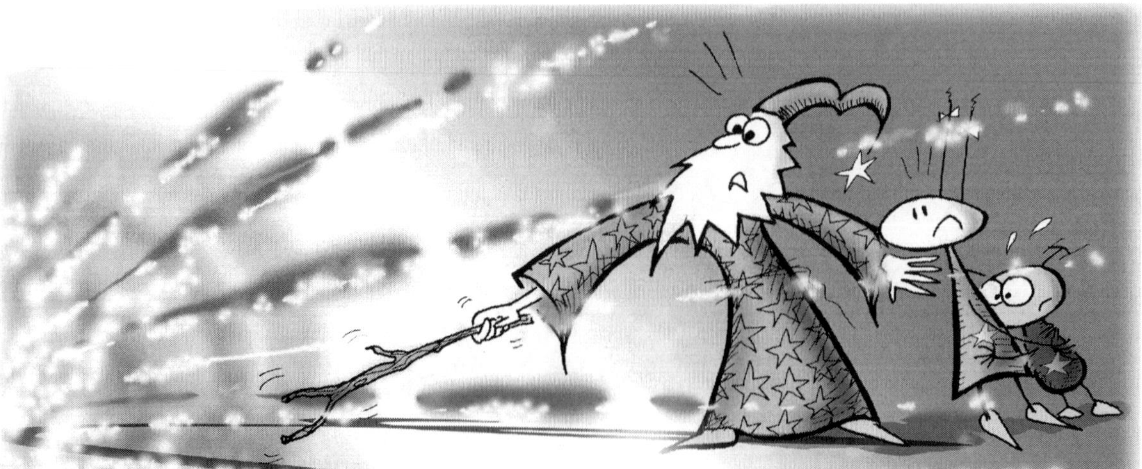

Chapter 26
What's this *really* about?

A million dollar prize, fiercely competitive individuals racing for a solution and eternal glory, stories of despair, elation, betrayal, paranoia and mental breakdown, possible implications for secret codes used in banking and national security (yes, prime numbers show up there too) – these are the ingredients which popular science journalism has tried to use to make exciting what most people would normally find the most utterly boring subject matter imaginable: egghead analytic number theorists beavering away over reams of indecipherable equations.

Although there's nothing wrong with attempting to cultivate a popular interest in these fascinating issues, I do feel that the excitement-based approach may overshadow something of real significance.

I certainly don't wish to condemn anyone involved in the quest for a proof of the RH. However, as with other branches of mathematics and science, there will unavoidably be researchers keeping secrets in the hope of bringing eventual glory and recognition to themselves above all others. Yes, this is perfectly understandable within the present cultural context, but it's also regrettable. Who knows what could be discovered if there were a complete, selfless cooperation between everyone involved!

Notice how all of the above ingredients of "exciting" journalism about the primes and the RH are concerned with the regrettable aspects of the situation – the results of human pride, greed and fear.

If we put all of these things aside, what do we see?

We're left with the question "why does the Riemann Hypothesis matter?" To answer *that*, we have to agree on the context of the discussion.

In the context of the lives of ordinary people – people who have no interest in or awareness of higher mathematics – the answer would most sensibly be "it doesn't". Life would go on much the same regardless of what anyone thinks, knows or proves about the zeros of the Riemann zeta function.

In the context of the lives of a miniscule fraction of the human population – the mathematicians (and a few physicists) who concern themselves with the RH – it's hugely important. Why? One reason we've already considered is that the truth or falsity of much of the mathematics they've constructed, or rely on, is linked to the truth of falsity of the RH. To know for certain that the RH is true would thus lead to enormous progress:

> "Right now, when we tackle problems without knowing the truth of the Riemann hypothesis, it's as if we have a screwdriver… But when we have it, it'll be more like a bulldozer." (Peter Sarnak)[1]

So there's the promise of technical advances and new tools which a proof of the RH could make available to mathematicians, allowing them to "bulldoze" the rockface of currently unknown mathematics. But also, as we've seen, some of the mathematicians involved in this quest have more personal, aesthetically-related reasons why the RH (and its possible truth or falsity) is important to them, as with Henry Iwaniec invoking Mother Nature:

> "Mother Nature has such beautiful harmonies, so you couldn't say that something like that is false."[2]

Resonating with my earlier remarks about "the music of the spheres", there's almost a sense that the Universe (or Mother Nature) would be somehow "letting us down" or "getting it wrong" if the RH were false. Marcus du Sautoy explains:

> "...Riemann's Hypothesis can be interpreted as an example of a general philosophy among mathematicians that, given a choice between an ugly world and an aesthetic one, Nature always chooses the latter."[3]

In characteristically poetic style, he here reveals something of how he imagines a post-RH mathematics:

> "A solution to the Riemann Hypothesis offers the prospect of charting the misty waters of the vast ocean of numbers. It represents just a beginning in our understanding of Nature's numbers. If we can only find the secret of how to navigate the primes, who knows what else lies out there, waiting for us to discover?"[4]

Finally, there's a third context in which we can ask this question "why does the RH matter?" This one transcends both the ordinary person unconcerned with mathematics and the analytic number theorist intent on proving new theorems. It's the context in which I believe David Hilbert was making his remark that the RH is the most important problem not only in mathematics but "*absolutely the most important.*"

To understand this, we must recall the extent to which the number system underlies our understanding and experience of "reality", as discussed in Chapter 1. Keeping that in mind, together with the centrality of the RH in our understanding of the number system, you can perhaps catch a glimpse of what Hilbert meant by his remark.

Although I can't pretend to know what Hilbert was thinking a century ago, and can't offer any kind of explanation as to how a solution of the RH might affect things at some fundamental level of our existence or perception, the final volume of the trilogy will be delving further into this "third context".

In the music-related quotations we saw earlier, there was an interesting cluster of words used in relation to the zeta zeros and the "harmonic" way in which they underlie the distribution of prime numbers: *arcane*, *secret*, *hidden*, *behind*, *mysterious* and *revealed*.

We'll pause for a moment and consider the common meanings and connotations of these words:

Arcane things are known only to the select few, those with some specialised knowledge, and are thereby hidden from everyone else.

Secrets are kept by someone from someone else.

Something is *hidden* if someone or something hides it from view.

If something is *behind* something else, there are two relevant connotations here:

(1) It's responsible for that thing, or somehow controlling it ("behind the scenes"). (2) it's at least partly obscured by that thing and thus cannot be properly seen.

Something is *mysterious* if it's difficult to explain or account for, of unknown origin, beyond human understanding, or in some way secret.

If something is *revealed* at some point, it must have previously been "concealed".

Read naively, all of this could create the impression that, at least prior to Riemann, "someone" or "something" was hiding the true nature of the prime numbers, keeping it secret, or concealing it from humanity. The question then arises as to "who" or "what" this would be.

There's a lot more of this in the popular literature.

Inspiring the name of this trilogy, the number theorist Don Zagier has talked about the primes in terms of "*inexplicable secrets of creation*".

Hermann Weyl has written of the "*the mystery of number, the magic of number*" being inherent in the distribution of primes[5]. Y. Motohashi has described the primes as "*very mysterious*"[6]. Martin Gutzwiller suggested that "*the zeta-function is probably the most challenging and mysterious object of modern mathematics, in spite of its utter simplicity.*"[7] Karl Sabbagh describes the Riemann Hypothesis as "*encapsulating a mystery at the heart of our number system.*"[8]

Here's du Sautoy again, from that 1998 article...

> "*The primes have fascinated generations of mathematicians and non-mathematicians alike, yet their properties remain deeply mysterious. Whoever proves or disproves the Riemann Hypothesis will discover the key to many of their secrets and this is why it… [is] the theorem for whose proof mathematicians would trade their soul with Mephistopheles.*"[9]

...and from his 2003 book:

> "*…despite their apparent simplicity and fundamental character, prime numbers remain the most mysterious objects studied by mathematicians.*"[10]

> "*The search for the secret source that fed the primes had been going on for over two millennia.*"[11]

> "*If we can only find the secret of how to navigate the primes, who knows what else lies out there, waiting for us to discover?*"[12]

In her *New Scientist* article on the RH, Erica Klarreich wrote:

> *"Proving the Riemann hypothesis won't end the story... what other jewels will we uncover when we dig deeper? Those who believe mathematics holds the key to the Universe might do well to ponder a question that goes back to the ancients: What secrets are locked within the primes?"* [13]

Who or what buried these jewels? Who's keeping the secrets? Who locked them inside the primes? Obviously these are rhetorical questions, but they raise the fascinating issue of subtle psychological forces at play, shaping the sentences these mathematicians and popularisers of mathematics commit to publication.

In the publisher's summary of John Derbyshire's book, we find this:

> *"Because Riemann was able to see beyond the pattern of the primes to discern traces of something mysterious and mathematically elegant at work – subtle variations in the distribution of those prime numbers... the successful solution to this puzzle would herald a revolution in prime number theory."* [14]

The "puzzle" is the RH. The word *puzzle* has the connotation of something someone has devised, the correct answer or solution being initially unknown to everyone else.

The following three quotations appear to be personifying the primes and the zeta function, as if they are somehow the keepers or creators of their own secrets:

> *"Littlewood's proof... revealed that prime numbers are masters of disguise. They hide their true colours in the deep recesses of the universe of numbers, so deep that witnessing their true nature may be beyond the computational power of humankind."* (Marcus du Sautoy) [15]

> *"...the prime numbers. Defined since antiquity, this key concept has yet to deliver up all its secrets – and there are plenty of them."* (Gérald Tenenbaum) [16]

> *"In 1985 there was a flurry of publicity for an announced proof of the Riemann Hypothesis... This announcement was premature, and the zeta function retains its secrets."* (Ian Stewart) [17]

Of course, the primes and zeta are not "active" entities which are able to "hide", "deliver" or "retain" anything – this kind of writing involves a kind of artistic license. But it adds to this impression of things being somehow hidden or concealed, secrets being kept, *etc.*

So, again, I ask who or what is responsible for the hiding, concealing, burying, locking and keeping of secrets?

If you were in a medieval European headspace, this would be straightforward – the answer would be "the Lord", "the Almighty" or just "God". However, we're not in medieval Europe, and many mathematicians inclined to use this language of "secrets" and "mysteries" may well be atheists, or even if not, might still disagree with answering the question in that way, or even with the appropriateness of the question itself.

In any case, I would argue that the use of this language and the fact that it "works" (it clearly resonates with the popular readership) reflects an unconscious tendency to see some sort of higher, or ultimate, creative power at work in these matters. This tendency seems quite compatible with Pythagoreanism (see Chapter 1) and could almost be described "neo-Pythagorean". We'll return to these issues before the end of Volume 3.

Arguably, the Riemann Hypothesis isn't really the issue here. Rather, it's *something else which the RH is pointing towards*.

We're about to see the mysterious fact that the number system can be harmonically decomposed eclipsed by a deeper mystery still. This will bring with it yet more surprise, suggesting strongly that *the heights of the zeta zeros are "vibrational frequencies" of something, and we have no idea what that something is.*

Don't worry if you're unsure about my use of the words "vibrational frequencies" at this point. You'll just have to accept for the time being that this situation is *very strange indeed*. In fact, in my decades-long pursuit of all things weird and wonderful, this would *easily* top the list. There is something "vibrating" "behind" or "at the root of" the system of counting numbers (this "something" having been represented by the music made by our trumpeting angels) and it has left its imprint as the zeros of the zeta function, only unearthed in the second half of the 19th century when Western mathematics had reached a suitable level of sophistication. Attempts to understand the Riemann zeros have led to a certain amount of new understanding, but they've also further deepened our bafflement, as we shall next see.

Chapter 27
a spectrum of vibrations

I may have created the impression a couple of chapters back that no one has a clue how to tackle the Riemann Hypothesis, that everyone involved is just waiting for a big new idea to come along. That's true to some extent, but there *is* a big idea which has gradually established itself in the mathematical community's imagination over the last few decades. Fairly recently, some remarkable evidence has emerged which strongly backs this idea up. It's called the *spectral interpretation* of the Riemann zeta zeros.

To understand this, we'll first have to look at *spectra* (the plural of "spectrum"). The word *spectrum*, to most people, evokes the image of a prism refracting a beam of light into a rainbow-coloured stripe (*see back cover*).

You might occasionally hear about a "spectrum" of something else, like political opinion on a particular issue, or the intensity of some disorder or disease. In that sense, it's meant as "a range of something which it's not obvious how you'd measure".

Like a sound wave, a beam of light is usually made up of a number of different sine waves combined. A prism uses simple geometry to split a beam of light up by spatially separating these waves according to their frequencies. Each frequency of light corresponds to a particular colour. Via *refraction* the prism geometrically reveals the frequencies (colours) of which the beam is made. If you shine pure white light through it, you'll get an unbroken rainbow stripe, the full spectrum of colours, because white light is made of all frequencies combined (all at the same level of intensity). If you use other types of light, you'll often get something very different. Astronomers use *spectroscopes* to analyse the light from distant stars. This also produces a spectrum, but rather than the continuous rainbow spectrum, it will consist of separate pieces – sometimes very narrow strips, sometimes wide bands. By looking at which colours are represented, it's possible to deduce what kinds of matter the star is made of since the material content of a star directly affects the light that it produces.

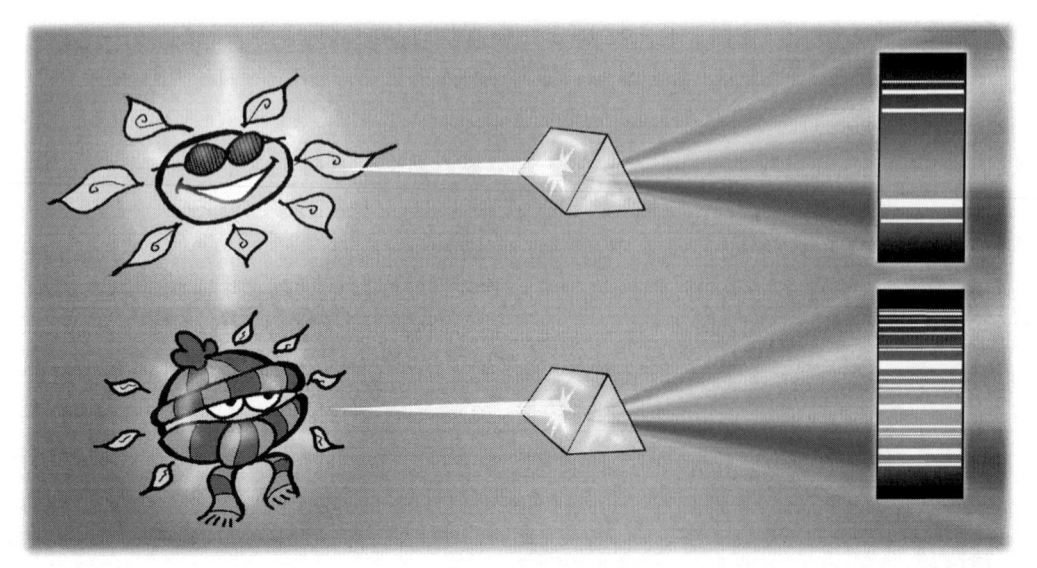

Hotter stars tend to consist mainly of hydrogen and helium, whereas cooler "red giants" contain a wider ranger of elements. These differences are evident in the spectra of the respective starlight.

As well as a spectrum of light, we can think in terms of a "spectrum of sound", since light and sound are both wave-based phenomena. What physicists call *white noise* is the sonic equivalent of white light. This can be heard in the whoosh of a waterfall or the static hiss of an untuned radio – all frequencies of sound are represented, at roughly the same level of intensity. A more musical sound would be made of a few notes or tones, each corresponding to a sine wave with a measurable frequency, a point on the spectrum of sound.

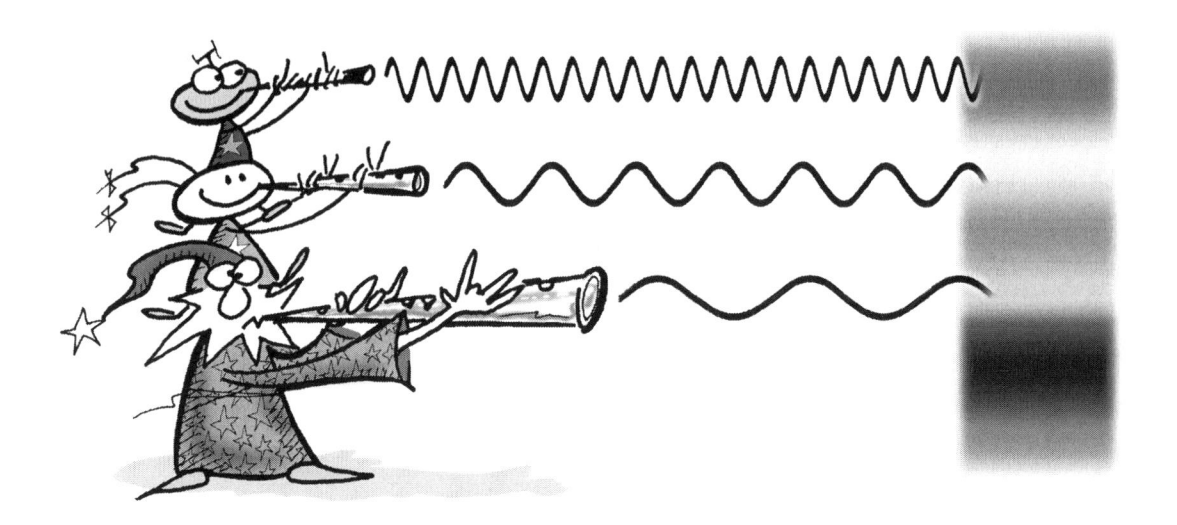

Acoustic scientists often perform *spectral analysis* on sound waves. The idea is to analyse the spectrum of a wave and deduce which frequencies it's built from. We saw an illustration in Chapter 13 of how composite sound waves are built up from sine waves. This is the reverse process – *decomposition*.

Even the entirely unmusical "clank" of a dropped saucepan could be, through spectral analysis, decomposed into a spectrum of specific sound frequencies (pure tones).

If you drop a squarish saucepan on a hard floor, it'll probably make a less pleasing "clank" than a more rounded one. Why is this? Some metal bowls can make beautiful ringing sounds. Some are even specially made for use in religious traditions (Tibetan prayer bowls and European church bells come to mind).

199

Let's consider why there's this difference in sounds between differently shaped saucepans. The *potential energy* of the elevated saucepan get converted by gravity to *kinetic energy* which brings about the collision of the saucepan with the floor, in this way getting converting into a kind of "vibrational" energy:

A whole range of waves then spread through the saucepan from the point of impact. Some of them "break" on the edges, or bounce off, fragmenting, cancelling each other out, and generally decaying very quickly. Collectively, these will just make a brief indistinct noise. However, there will be a few waves which, due to the exact size and shape of the saucepan, will be able to sustain themselves for some time – the combination of the wave's geometry and that of the metal object it's passing through being just right. These *standing waves* will make up any tones which you can hear persisting in the "clank".

Bells and various wind-based musical instruments have the shapes they do for very good reasons. Through centuries of trial-and-error and experimentation, instrument design has incorporated certain principles of "spectral geometry".

The easiest way to see how this works is to consider the one-dimensional equivalent – the vibrating string. When you pluck or twang a guitar string, a whole range of waves are set off, but only a select few are able to persist. Those are the waves whose halved wavelength is either the length of the string itself, half that length, a third, quarter, fifth, sixth, or any similar fraction:

Waves of any other wavelength rapidly fade, whereas the "standing waves" just described are able to sustain themselves for considerably longer – they give us the "harmonic frequencies" of the string.

Now try to imagine a much more complicated version of the same thing, where waves are spreading out across a (possibly curved) surface. It's generally quite difficult to visualise this. But in the 1780s it was discovered by Ernst Chladni that if you sprinkle sand on a flat metal plate and then apply a vibration of constant frequency to the plate (originally with a violin bow, these days with an electronic sound generator), distinct geometric patterns emerge in the sand. Because the standing waves are more "active" in some regions of the plate than others, their vibrations shift the sand away from these regions so that it collects in those areas where the the waves are "least active" (lines and curves which are analogous to the *node points* seen in the previous illustration). These *Chladni figures* reveal something visually about the standing waves which are supported by the shape of the plate being used (below these all being squares).

Sometimes (as with musical bowls and bells), a sound consists of one very clearly dominant tone with perhaps a couple of extra "harmonics" just barely audible. Your average saucepan, on other hand, produces a blurry, dissonant chord – a cluster of tones all ringing simultaneously. Each of these tones can be thought of as a frequency of sound and given a number which measures that frequency. For example, a bell or bowl ringing out with a concert pitch "A" note is vibrating at a frequency of 440 vibrations per second (*Hertz*). The higher the frequency, the higher pitched the note becomes (human hearing is in the range of about 20–20000 Hz). The collection of all vibrational frequencies which persist in the "clank" of a saucepan could be called the (acoustic) spectrum of that saucepan. So, to put it very crudely, the spectrum of the saucepan is a collection of numbers which describes what it would sound like if you were to drop it or hit it.

In fact, there's a branch of mathematical physics which is concerned with "what things would sound like if you could hit them". *Spectral geometry* consists of techniques by which you can study the geometry of an object and deduce from it facts about the object's spectrum. Going in the opposite direction, there's the study of *inverse spectral problems* which involves starting with the spectrum of an object ("what it sounds like") and deducing facts about its geometry ("what it looks like"). Physicists study "vibrating systems" (in the acoustic sense, but also in more general contexts) and their spectra of *eigenvalues* (frequencies, in a very general sense), all of this being expressed in the form of precise equations like this:

$$\Delta f(t,\xi) = \sin^{2-n} t \frac{\partial}{\partial t}\left(\sin^{n-2} t \frac{\partial f}{\partial t}\right) + \sin^{-2} t \Delta_\xi f$$

Returning from acoustics to astronomy (where we started our discussion of spectra), a star could be thought of as a "vibrating system", light and its other forms of energetic output being vibrational in nature. The spectral analysis of the star's light will provide information not about the *geometry* of the star, but about its physical constituents.

Spectral analysis and spectral geometry in these contexts are part of what's known as *applied mathematics*, the systematic use of mathematical techniques to solve problems in the physical world.

"Pure mathematicians" have taken these ideas on from applied mathematics and learned to use the same sort of thinking to explore wholly abstract, nonphysical "structures" in which they're interested. These structures (or "objects", or "entities") can be described with geometry, but *not* the simple geometry which most people are familiar with. By calculating the spectrum of something like an "Enriques surface", a "Klein bottle"[1] or a "seventeen-dimensional hypersphere"[2] (none of which could be constructed in physical reality), a mathematician is finding out, in some sense, "what it would sound like if you were able to make one and hit it".

I must stress, though, that within applied mathematics, the study of spectra isn't limited to just light and sound. There's also a spectrum of *electromagnetic energy*. This includes X-rays, radio and television frequencies, microwaves, infra-red and ultraviolet radiation *and* the whole of the visible light spectrum (visible light is just electromagnetic radiation that human eyes can detect, that is, one small piece or *band* of the overall electromagnetic spectrum):

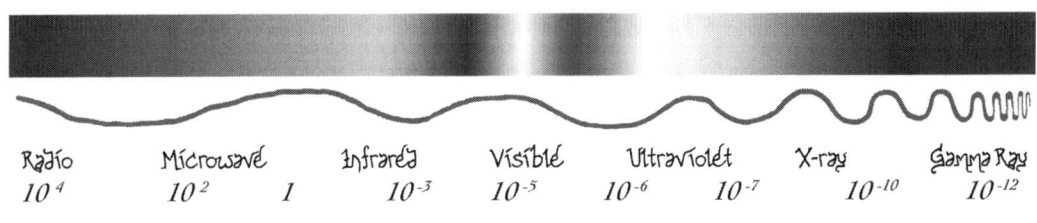

Radio	Microwave	Infrared	Visible	Ultraviolet	X-ray	Gamma Ray
10^4	10^2 1	10^{-3}	10^{-5}	10^{-6}	10^{-7} 10^{-10}	10^{-12}

The electromagnetic spectrum (wavelengths given in metres).

Spectra of all kinds show up across the sciences – astronomy, optics, acoustics and electromagnetic theory, as I've described, but also in atomic physics and mechanical engineering. In fact, whenever you're studying something that vibrates, you'll find a spectrum.

Physicists, and sometimes mathematicians, deal with various types of *dynamical systems*. Think of the difference between a still photographic image and the moving image on a cinema screen. The geometry you're familiar with (studying points, lines, shapes in space) is like the still image. Dynamical systems theory, on the other hand, would correspond to the film – geometry in motion. With a dynamical system, there's always some sense of *time passing* (even if this "time" is a mathematical abstraction), during which "stuff happens in space" (this "space" also possibly being a mathematical abstraction) according to some mathematical laws. A set of balls in motion on a billiard table, the solar system and a simple pendulum are all examples.

Descending into the world of atomic and subatomic physics, we find dynamical systems which have spectra of *energy levels* (energy levels and frequencies are very closely related in these realms of physics)[3].

Sometimes mathematicians study "abstract" dynamical systems with no relation to physical reality, but these can then go on to become part of the physicist's toolkit for creating mathematical models of *actual* physical systems whose spatial configurations "evolve" through time (a yo-yo, an atom, a solar system).

Some of what subatomic physicists do involves predicting, observing and attempting to account for various aspects of the spectra of different types of subatomic physical systems. Such systems could be individual particles or groupings of particles, subject to certain physical forces and environments, and interacting according to the (entirely mathematical) laws of physics at that scale. Certain facts can be deduced about the system by studying the appropriate characteristics of its spectrum.

As a result of this sort of activity, mathematicians and physicists have become interested in finding identifying features ("fingerprints") in certain types of spectra which can be used to determine the type of vibrational system which produced them. This is perhaps comparable to an entomologist, identifying a certain species of insect by its unique sound. A particular buzz (which would sound like hundreds of other buzzes to the uninitiated ear) could in some cases be linked to a species-of-origin by a skilled ear. Even though each individual insect of that species would produce a slight variation on the same buzz, they would all have something in common, something which could be traced back to *common features in their spectra*. In a similar sort of way, a spectrum of frequencies (or energy levels) can, in some circumstances, be traced back to a certain kind of vibrating physical system (whether a dropped saucepan of a particular shape, a violin string or a pair of atoms interacting in a particular magnetic field).

As I've mentioned, there are areas of mathematics with no known relevance to physics, *but which still involve spectra of various types.* It's not all about Klein bottles and seventeen-dimensional hyperspheres, though – these still have *something* in common with familiar geometrical objects. The idea of a spectrum can be entirely "abstracted" away from both physics and geometry.

There are some very commonly used objects throughout higher mathematics called *matrices* (the plural of "matrix"). Matrices are rectangular blocks of numbers of one sort or another – real, rational, integer, complex, whatever.

The individual numbers are called the *entries* of the matrix. Matrices can (if "compatible" in a particular sense) be added together, subtracted, multiplied and have elaborate mathematical operations performed on them. They generate a whole mathematics of their own (there's something called *matrix algebra*, for example).

206

The square ones (those with the same number of rows and columns) are especially important. Associated with each of these is a set of numbers called its *eigenvalues*. In general, the number of eigenvalues of a square matrix is the same as its number of rows (or columns), so a 5 × 5 matrix, for example, generally has 5 eigenvalues, while 117 × 117 matrices will generally have 117 eigenvalues each [4].

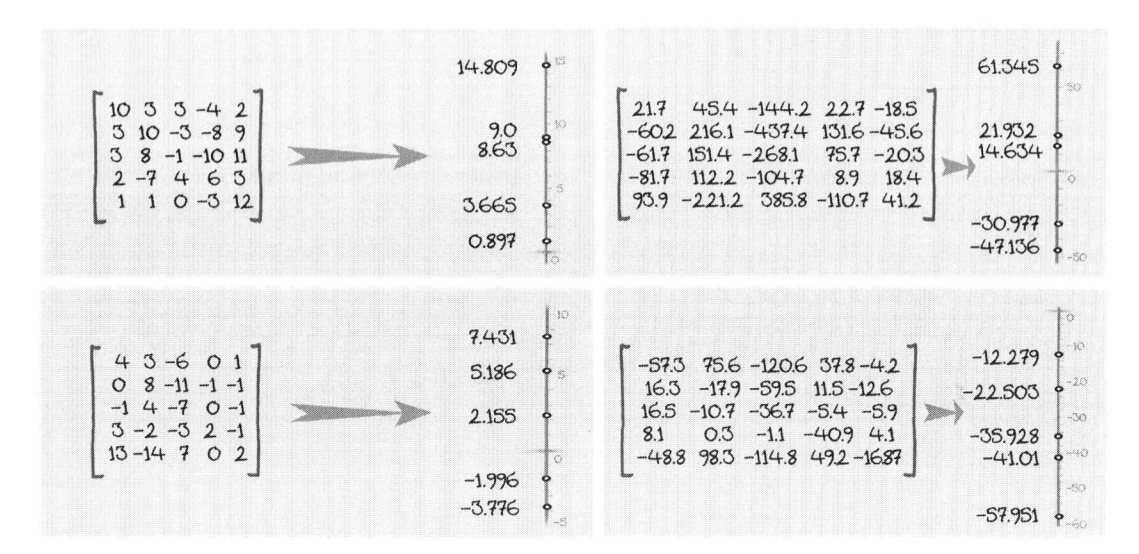

Each of the 5×5 matrices shown has five eigenvalues. These are all real numbers – the fact they're displayed on a vertical axis has nothing to do with imaginary numbers.

These sets of numbers are also called *spectra*. There's a good reason for this. In certain areas of physics (such as quantum mechanics), the spectrum of a vibrating system can be found by calculating the eigenvalues of a matrix which is mathematically linked to the structure of the system. But matrix eigenvalues are still sometimes collectively described as a "spectrum" even if they're just some numbers chalked up on a blackboard with no relevance at all to anything physical.

How you calculate the eigenvalues of a matrix need not concern us. You'll just have to accept that there's a well-established mathematical process for doing this and that the spectrum of eigenvalues follows deterministically from the entries of the matrix. Because a matrix can, in some situations, be related to a vibrating physical system (with the matrix's spectrum of eigenvalues corresponding to the system's spectrum of vibrational frequencies), we can loosely think of the eigenvalues of *any* square matrix as representing something like its "resonant frequencies". If those words mean nothing to you, then it's probably best to think (again, loosely) of the spectrum of eigenvalues representing "the sound of a matrix if you could make one and hit it"!

Physicists whose work involves matrices are able to make use of the intricate matrix mathematics developed by pure mathematicians. Certain types of "large" square matrices (which have "a lot" of eigenvalues) are relevant to subatomic physics. Physicists are interested in studying the corresponding physical spectra and looking for patterns therein, trying to get a deeper understanding of the vibrating systems which produce them. Meanwhile, mathematicians have been looking at different classifications of matrices (based on how their various entries relate to each other). These various "families" of matrices produce spectra which, although all differing individuals, show a common pattern. There has been some productive interaction between mathematicians and physicists on this matter in recent years, so certain types of matrices have now been firmly linked to certain types of subatomic systems (based on correspondences in their spectra).

Here are a few different types of spectra, represented as horizontal *spectral lines*, the vertical scale representing the total spectrum (frequencies of light, sound, electromagnetic radiation, mechanical vibration, *etc.*):

Some spectra associated with (left to right): iron atoms, matrices with random entries, a dynamical system known as "Sinai billiards", the Sun and a biological system.

There's another class of mathematical objects, called *operators*, which also have spectra of eigenvalues. Both mathematicians and physicists use various types of operators in their work. Sometimes, these can be expressed as (square) matrices. 2×2 matrices can be thought of as simple "geometric" operators — you can use them to carry out various geometric "operations" on a two-dimensional space (a plane) such as rotation, reflection across a line, magnification and shrinking. (This involves matrix multiplication and the representing of points in the plane by 2×1 matrices, but it's not something that we'll need to use.)

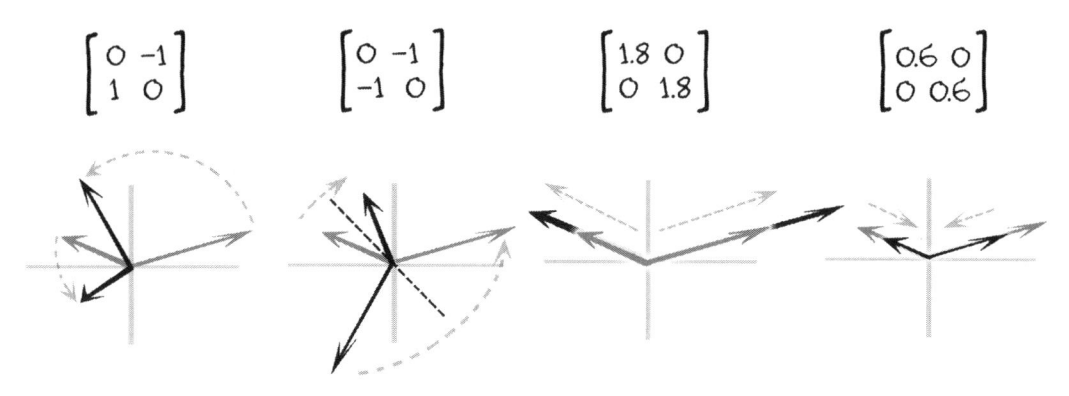

$$\begin{bmatrix} 0 & -1 \\ 1 & 0 \end{bmatrix} \quad \begin{bmatrix} 0 & -1 \\ -1 & 0 \end{bmatrix} \quad \begin{bmatrix} 1.8 & 0 \\ 0 & 1.8 \end{bmatrix} \quad \begin{bmatrix} 0.6 & 0 \\ 0 & 0.6 \end{bmatrix}$$

Left to right: rotation, reflection across a line, magnification (dilation), contraction (shrinking)

Similarly, 3×3 matrices can act as geometrical operators on three-dimensional space. But operators don't often have this kind of straightforwardly "geometrical" quality. An operator is a *very* general kind of mathematical entity, basically "something that does something to something else" (a crude but fairly accurate description).

If an operator can be represented by a matrix, then its eigenvalues will simply be the eigenvalues of that matrix. If it can't, then a more abstract definition of "eigenvalues" is required, but this will still have a strong mathematical connection to the matrix-related definition.

Some operators arise directly from physics, some have been developed by pure mathematicians and later found to be of use in physics, while some are purely mathematical concepts, they and their spectra seemingly existing without any connection to physical reality. Operators tend to have spectra involving *infinitely many* eigenvalues, and so it's sometimes useful to think of them as resembling "infinitely large square matrices".

In the course of these first two volumes, we've encountered some things which might have seemed surprising to you. They've certainly surprised quite a few people over the years. But we now come to what must surely be the most shockingly strange discovery, mathematical or otherwise, of the modern age:

> *The "heights" (imaginary parts) of the Riemann zeta zeros are almost certainly a spectrum of eigenvalues of some unknown vibrating system.*

This is the "spectral interpretation" of the Riemann zeta function. Formally, it's still an unproven conjecture, but the body of evidence (much of it due to Andrew Odlyzko's prolific computations) is now so overwhelming that Peter Sarnak, unquestionably a leading expert on the matter, has gone as far as to say that the zeros "*are absolutely, undoubtedly 'spectral' in nature*" [5].

So, it's now almost universally accepted that the zeta zeros' heights correspond to the spectrum of some kind of operator. The details of this operator, however, remain a profound mathematical mystery.

In simple terms, as Marcus du Sautoy puts it [6]:

> "*We have all this evidence that the Riemann zeros are vibrations, but we don't know what's doing the vibrating.*"

We may not know what it is that's "doing the vibrating", but we can still say quite a lot about it. And what we can say has some profoundly weird implications! This shall be where we take up the story in the third and final volume, *Prime Numbers, Quantum Physics and a Journey to the Centre of Your Mind*.

End of volume two

notes

A RE-INTRODUCTION [pages 1–3]

1. Bernhard Riemann, "Über die Anzahl der Primzahlen unter einer gegebenen Grösse", *Monatsberichte der Berliner Akademie* (November 1859) [English translation available at http://tinyurl.com/33cfvb8]

2. "Object" might seem like a strange word to use here – it's hard to choose a nontechnical word to describe some of the things mathematicians work with. It's certainly not an object in the sense that you could pick it up or put it in your pocket. "Thing" might actually be a better word, but some people would find that too vague. "Entity" would be about right, but then that sounds too formal to a lot of people. So we'll stick with "object".

CHAPTER 16 [pages 5–28]

1. Remember, numbers are being *represented* as points on a line in this explanation, but it's important to remember that *this is just a helpful representation*, not an attempt to claim that points on a line and numbers are somehow identical. Also, "points" are here considered to be infinitely small, not to be confused with visible dots, which have a finite size (however small they are).

2. F. von Lindemann, "Über die Zahl π", *Mathematische Annalen* 20 (1882) pp. 213–225

3. An outline of Fourier's proof (by contradiction) that e is an irrational number can be found at http://en.wikipedia.org/wiki/Proof_that_e_is_irrational [http://tinyurl.com/58h7v6]

4. In fact, there are more. You might question how one infinite amount of something can be "bigger" than another infinite amount of something, but I assure you that there is a meaningful sense in which it can! If you're interested, look up "transfinite numbers".

5. It's possible to add or multiply a pair of irrational numbers and get a result which is not an irrational number. For example, the (irrational) square root of 2 multiplied by itself produces the rational number 2. So the set of irrational numbers is not "closed" with respect to addition and multiplication, and thus can't be considered a number system.

6. The definition of a completion involves *equivalence classes* of *Cauchy sequences*. A Cauchy sequence is an infinite sequence (in this case of rational numbers) whose elements tend to cluster together in a very particular way, but to define *that* precisely requires the choice of what's called a

metric as well as the notion of a *limit*. To explain equivalence classes, I would have to first explain what an *equivalence relation* is. To keep "unpacking" these concepts would take up more and more space, hence my need to occasionally be vague about them!

7. It's all to do with "homing in" on the holes in $\mathbb{Q}$ by finding sequences of numbers that are in $\mathbb{Q}$ and which take you ever closer to (but never *into*) a hole. This relies on the idea of distance between points and *there's more than one way to define distance*. The obvious "ruler"-style distance leads you to the usual number system on the line, the one known as $\mathbb{R}$. But it's possible, using a fixed prime number, to define a kind of distance between a pair of numbers in $\mathbb{Q}$ in terms of the role of that prime number in the factorisations of certain counting numbers related to the pair. You take the usual "ruler"-style distance between them, express it as a fraction (for it will itself be a rational number) then look at the numerator and denominator of this fraction and determine how many times the fixed prime number divides into each. Based on that, a *p*-adic distance can be defined. The study of *p*-adic number systems is, for this reason, closely linked to number theory.

8. Actually, if negative numbers are involved, then there is a sort of 180° rotation. This is seen taking place in the illustrations on p. 24.

9. Think about it like this: At a time when you're completely broke (your money = 0), you suddenly find you have received eight fines for overdue library books. Each one is going to cost you £2. This means that you're in debt $8 \times 2 = 16$ pounds to the library. Suddenly your money = -16. But you plead your case with the librarian that five of these eight fines are unfair because the books were mislabelled, suggesting longer loan periods. Your plea is accepted, so you're able to get rid of five £2 fines, or equivalently, *you gain -5 two pound fines*. This would add -5×-2 to your money. And what happens to your money? You are £10 better off, clearly, so you go from -16 to -6 (you now have 3 debts of -2 pounds = $3 \times -2 = -6$ pounds). So gaining -5 amounts of -2 pounds results in gaining 10 pounds, that is, $-5 \times -2 = 10$.

10. Taken from H. Guenancia, "Toric plurisubharmonic functions and analytic adjoint ideal sheaves" (preprint, November 2010)

11. The two fractals pictured are examples of what are known as *Julia sets*. These have been known about by mathematicians since the early 20th century, but (unsurprisingly) no one was able to draw one by hand, so mathematics had to wait for computer graphics to arrive.

12. In the physicist Roger Penrose's book *The Emperor's New Mind* (Oxford University Press, 1989) these $\mathbb{C}$-inhabiting fractals are presented as a kind of evidence for a "Platonist" interpretation of "mathematical reality" – that it has some kind of independent existence beyond the realm of human thought (the book itself looks at issues concerning artificial intelligence and human consciousness, covering a wide range of topics in the process). However, the Platonist view is strongly contested by other thinkers from the "constructivist" school of mathematical philosophy.

I once heard a quite convincing argument from a German postgraduate student (although I wasn't convinced) that there's nothing remarkable or beautiful about fractals, *it just seems that way* because of how our minds are structured.

CHAPTER 17 [pages 29–42]

1. Note that we could have also made a spiral of base 1 (circle of radius 1) and everything would remain consistent. But if this confuses you, it's probably best not to dwell on it.

CHAPTER 18 [pages 43–66]

1. Going back to our definition, you'll find that the Euler zeta function transports the value 2 to $1 + 1/4 + 1/9 + 1/16 + 1/25 + \cdots$. This sum can be proved to equal π times itself divided by 6. This provides one way to calculate π (although not a very efficient one).

2. The idea of choosing two counting numbers at random is a bit of a subtle one, so we need to be quite precise about this. Random selection in probability theory is initially defined in terms of choosing from a finite collection (a single playing card from a pack of 52, for example). So we begin by discussing the probability that two randomly chosen counting numbers between, say, 1 and 1000 have no prime factors in common. This notion can be made rigorous and a straightforward calculation will yield a quantity reasonably close to 0.608. By then looking at randomly chosen counting numbers between 1 and 10000, 1 and 100000, *etc.*, we get different probabilities, but as our "range of selection" grows, these will get closer and closer to the exact value 0.6079271... The mathematical concept of a *limit* can then be applied, the idea being that the exact value can be approximated as closely as is desired simply by making the range of selection large enough.

3. See Chapter 8, "In Which Christopher Robin Leads an Expotition to the North Pole", in A.A. Milne, *Winnie-the-Pooh* (Methuen & Co., 1926).

4. This actually *will* be a logarithm, but of a different base. Recall that the base-*e* spiral gives "base *e*" logarithms, also known as "natural logarithms". Other equiangular spirals give logarithms of other bases.

CHAPTER 19 [pages 67–80]

1. From MSRI video lecture "Random Matrix Models" (UC Berkeley, January 1999) available at http://www.archive.org/details/lecture11493 [http://tinyurl.com/cuy2yc6].

2. The gamma function is an extension of the *factorial function* from the natural numbers to the complex plane. The factorial of a positive integer n, written $n!$, is defined to be $n\times(n-1)\times\cdots\times2\times1$, so $7! = 7\times6\times5\times4\times3\times2\times1 = 5040$, for example. See Appendix 12 for more detail.

3. "...it is very probable... Certainly, it would be desirable to have a rigorous proof of this proposition; however, I have temporarily put aside my search for this after some brief and unsuccessful attempts, since it appears to be unnecessary for the immediate goal of my investigation." [excerpt from "Über die Anzahl der Primzahlen unter einer gegebenen Grösse", *Monatsberichte der Berliner Akademie* (November 1859), my translation]

4. Although if you have a look around online, you'll find a few people who think they *have* proved it. None of the proofs in circulation have received the approval of the mathematical community at large. There will be more about this in the Chapter 25.

5. The fact that there are none to the right of the critical strip can be deduced from the Euler product formula. Recall that the zeta function can be defined there in terms of a procedure which involved multiplying infinitely many complex numbers together. It was well known in Riemann's time that such an "infinite product" can only equal zero if one or more of the complex numbers involved is zero. And if you go back and look at the procedure (which we saw in terms of little arrows inscribed within spirals), you'll see that none of them *can* equal zero. Using the functional equation, it then immediately follows that no point to the left of the critical strip, *apart from the negative even integers*, can be a nontrivial zero. For a more thorough account of this, see Appendix 11.

6. It says that the number of nontrivial zeros above the real axis and below a certain "height" is given by approximately the following formula. Take your height and divide it by $2\times\pi$ (that's about 6.28). Call this the "rescaled height". Now multiply the rescaled height by its own logarithm, and subtract the rescaled height from the result. Here's an example: To find the approximate number of nontrivial zeros below 1000 in the upper half of the critical strip, we divide 1000 by $2\times\pi$ to get about 159.155. The logarithm of 159.155 is about 5.069, so we multiply 159.155 by 5.069 (to give 806.896...) and then subtract 5.069, giving the final answer 801.826.... So there are approximately 801.826... nontrivial zeros in this region according to this approximate logarithmic formula. The actual number is 649, so we have only 76.5% accuracy at this "height" – not particularly good. The higher you go, though, the more accurate the formula becomes (coming as close to 100% accuracy as you like if you're prepared to look high enough).

CHAPTER 20 [pages 81–97]

1. It's usually known as the *Riemann–von Mangoldt explicit formula* since Hans von Mangoldt developed it some decades after Riemann published his original version.

2. This spiral Ferris wheel visualisation, it was originally noted, only produces that part of the spiral wave to the right of 1. The other section, between 0 and 1, isn't actually that important when you're looking at trying to reconstruct the prime count staircase – there are no primes between 0 and 1! But to build it, remember, we just visualise the bicycle starting off at 1 but facing to the left, following the same rule (so it will be *slowing down* this time, approaching 0 but never quite reaching it), with the Ferris wheel turning in the opposite direction.

3. This is necessary because there's a distinction between "frequency" and "angular frequency" for sine waves. The latter is $2 \times \pi$ times the former. In the spiral wave analogy, the zeta zero heights represent *angular* frequencies, so to get ordinary frequencies, we divide by $2 \times \pi$.

4. If we were to create the second wave by rotating the spiral so that the carriage is on the positive horizontal – rather than the vertical – axis, as with the first, then the initial amplitude of the first wave would be *exactly* that many times greater than the second wave's initial amplitude.

CHAPTER 21 [pages 99–117]

1. E. Bombieri, "Prime Territory: Exploring the Infinite Landscape at the Base of the Number System", *The Sciences*, September/October 1992, p. 36

2. Of course we can never rule out that someone did but kept quiet about it.

3. This has been slightly adapted from a formula that can be found in A.P. Guinand, "A summation formula in the theory of prime numbers", *Proceedings of the London Mathematical Society* (2) 50 (1948) 107–119

4. Fourier analysis involves breaking waveforms or signals down into their component sine waves. Unlike the situation with the "primeness count deviation", there won't usually be such a "clean" separation into a set of distinct waves – instead, different frequencies are represented with varying "power". The *Fourier transform* of a function is a second function showing the relative powers of each frequency found underlying the first. When two functions are each other's Fourier transform, we have the simplest instance of a Fourier duality (although the term has a more general usage, that's the basic idea).

5. G. Tenenbaum and M. Mendès France, *The Prime Numbers and Their Distribution* (AMS, 2000) p. 1

6. G.H. Hardy and J.E. Littlewood, "The zeros of Riemann's zeta-function on the critical line", *Mathematische Zeitschrift* 10 (1921) pp. 283–317

7. J.B. Conrey, "More than two fifths of the zeros of the Riemann zeta function are on the critical line", *Journal für die reine und angewandte Mathematik* 399 (1989) pp. 1–16

8. J.P. Gram, "Note sur les zéros de la fonction $\zeta(s)$ de Riemann", *Acta Mathematica* 27 (1903) pp. 289–304

9. R.J. Backlund, "Sur les zéros de la fonction $\zeta(s)$ de Riemann", *Comptes Rendus de l'Académie des Sciences, Paris* 158 (1914) pp. 1979–1982

10. J.I. Hutchinson, "On the roots of the Riemann zeta-function", *Transactions of the American Mathematical Society* 27 (1925) pp. 49–60

11. This unfinished machine is not to be confused with the more famous (but abstract, non-physical) "Turing Machines" studied by computational theorists. Despite arguably contributing more to the defeat of Nazi Germany than any other British citizen (he was centrally involved in decrypting the German "Enigma" code), Turing was persecuted by the British authorities for his sexuality, eventually driving him to suicide.

12. A. Odlyzko, "The 10^{22}-th zero of the Riemann zeta function and 175 million of its neighbors" (unpublished, 1989) [available at http://tinyurl.com/35fbckh]

13. A. Odlyzko, "The 10^{22}-nd zero of the Riemann zeta function", *Contemporary Mathematics* 290 (2001) pp. 139–143

14. See http://www.zetagrid.net/ [also archived at http://tinyurl.com/3xbph23].

15. Note that the mirroring pair to the left of the critical line will produce a pair of spiral waves with amplitude growth rate *less than* ½. Rather than eventually dwarfing all of the others, these waves will eventually be dwarfed by them.

16. E. Bombieri, "Prime Territory: Exploring the Infinite Landscape at the Base of the Number System", *The Sciences*, September/October 1992, p. 36

17. K. Sabbagh, *The Riemann Hypothesis: The Greatest Unsolved Problem in Mathematics* (Farrar, Straus and Giroux, 2003), p. 36

18. E. Bombieri, "Prime Territory: Exploring the Infinite Landscape at the Base of the Number System", *The Sciences*, September/October 1992, p. 36

CHAPTER 22 [pages 119–137]

1. M.C. Gutzwiller, *Chaos in Classical and Quantum Mechanics* (Springer, 1990) p. 308

2. Taken from Maslanka's now-defunct homepage http://www.oa.uj.edu.pl/~maslanka/ [archived as http://tinyurl.com/3a3w5kk].

3. R. Bellman, *A Brief Introduction to Theta Functions* (Holt, Rinehart and Winston, 1961) p. 30

4. A. Elon, *Jerusalem: City of Mirrors* (Fontana, 1989) p. 32

5. W.D. Smith, "Cruel and unusual behavior of the Riemann zeta function" (unpublished, 1998) [available at http://tinyurl.com/2vozykt]

6. From MSRI video lecture "Random Matrix Models" (UC Berkeley, January 1999) [available at http://www.archive.org/details/lecture11493]

7. Carl Gauss, the so-called "Prince of Mathematicians", is quoted as having said *"Mathematics is the queen of the sciences and number theory is the queen of mathematics."* [W.S. von Waltershausen, *Gauss zum Gedächtniss* (1856)]

8. About ten years ago, my friend Stevie came back from India with a "Shaktiman" T-shirt and explained that Shaktiman was India's first homegrown TV superhero. Despite being clearly modelled on classic American comic book superheroes, the character had been very much adapted to Hindu culture. And, most interestingly, Stevie had noticed that rickshaw drivers, known to emblazon their rickshaws with stickers representing their family, regional or otherwise favoured Hindu gods and goddesses, *had also started to include Shaktiman stickers*. TV superheroes, though, are more popularly accessible than analytic functions of a complex variable, so it might be a while before the Riemann zeta function stickers start appearing.

9. F. Galton, *Natural Inheritance* (Macmillan & Co., 1889) p. 66. Be warned that Galton was also an enthusiastic proponent of eugenics.

10. Nassim Nicolas Taleb, *The Black Swan: The Impact of the Highly Improbable* (Random House, 2007). Taleb is a fierce critic of the lazy and inappropriate use of the Gaussian distribution in economics and the social sciences, and what he calls the tendency to "see bell curves everywhere". One chapter of his book is entitled "The Bell Curve, That Great Intellectual Fraud". He's not questioning any of the accepted mathematical theory surrounding this object, though, just its misapplication to the physical world.

11. From Johan Andersson's website, hosted by Stockholm University's mathematics department 2001–6. It's archived here: http://tinyurl.com/3yg749r

12. P. Collins, "A Deeper Significance: Resolving the Riemann Hypothesis" [unpublished, available at http://tinyurl.com/2undeqn]

13. P. Erdős and M. Kac, "The Gaussian law of errors in the theory of additive number theoretic functions", *American Journal of Mathematics* 62 (1940) pp. 738–742.

14. Nassim Nicolas Taleb, *The Black Swan: The Impact of the Highly Improbable* (Random House, 2007)

15. T. Gowers, *Mathematics: A Very Short Introduction* (Oxford University Press, 2002) p. 121

16. G.H. Hardy and J.E. Littlewood, "Some problems of 'partitio numerorum' III: on the expression of a number as a sum of primes", *Acta Mathematica* 44 (1922) p. 37

17. See P.D.T.A. Elliot, *Probabilistic Number Theory II: Central Limit Theorems* (Springer, 1980).

CHAPTER 23 [pages 139–149]

1. B. Riemann, "Über die Anzahl der Primzahlen unter einer gegebenen Grösse", *Monatsberichte der Berliner Akademie* (November 1859) pp. 671–680 [my translation]

2. C. Reid, *Hilbert* (Springer, 1996) p. 92 [italics mine]

3. R. Bellman, *A Brief Introduction to Theta Functions* (Holt, Rinehart and Winston, 1961) p. 33

4. E.T. Bell, *Men of Mathematics* (Simon & Schuster, 1937) p. 538

5. E. Bombieri, "Prime Territory: Exploring the Infinite Landscape at the Base of the Number System", *The Sciences*, September/October 1992, p. 36

6. K. Devlin, *Mathematics: The New Golden Age* (Columbia University Press, 1999) p. 199

7. K. Sabbagh, *The Riemann Hypothesis: The Greatest Unsolved Problem in Mathematics* (Farrar, Straus and Giroux, 2003) p. 36

8. *ibid*, p. 219

9. *ibid*, p. 222

10. *ibid*, p. 36

11. M. du Sautoy, *The Music of the Primes: Why an Unsolved Problem in Mathematics Matters* (HarperCollins, 2003) p. 18

12. E. Bombieri, "Prime Territory: Exploring the Infinite Landscape at the Base of the Number System", *The Sciences*, September/October 1992, p. 36

13. E. Bombieri, from "The Riemann Hypothesis", in *The Millenium Prize Problems*, editors J.A. Carlson, A. Jaffe and A. Wiles (AMS/Clay Maths Institute, 2006)

14. K. Sabbagh, *The Riemann Hypothesis: The Greatest Unsolved Problem in Mathematics* (Farrar, Straus and Giroux, 2003) p. 268

15. *ibid*, p. 269

16. *ibid*, p. 189

17. *ibid*, p. 246

CHAPTER 24 [pages 151–167]

1. Note that this isn't the graph of a function since there are vertical lines which intersect it in more than one place (the graph of a function will never involve more than one point above/below any given number on the number line).

2. These sets of numbers are not actually "series" in the formal mathematical sense (which involves infinite sums). John Farey, Sr. was a geologist and amateur mathematician who published a letter in a philosophical journal about these numbers in 1816. It turns out that the mathematician Charles Haros had previously published something about them in 1802, but for historical reasons, they've ended up linked to Farey's name.

CHAPTER 25 [pages 169–187]

1. Someone did once name something of mine the "Watkins Objection" though. It's a very strange story (you can find out more at `http://www.secretsofcreation.com/2012.html` if you're interested).

2. *L*-functions are defined in terms of infinite sums of complex numbers, rather like what we did on pages 58–59, except that now each of the arrows in the sequence corresponding to the counting numbers gets multiplied by some complex number. This sequence of complex numbers characterises the *L*-function in question. If we use the sequence $1, 1, 1, 1, 1, \ldots$ then we just end up with a special case: the *L*-function known to us as the Riemann zeta function.

3. In 1933, Helmut Hasse proved a "Riemann Hypothesis" associated with "elliptic curves". André Weil extended this by proving the Riemann Hypothesis for a much wider class of "algebraic curves" in 1948. Pierre Deligne then proved a further generalization in 1973.

4. J. Brian Conrey, "The Riemann Hypothesis", *Notices of the AMS* (March 2003) p. 347

5. A. Deitmar, "Panorama of zeta functions" (unpublished paper, 2005) p. 12

6. H.M. Edwards, *Riemann's Zeta Function* (Dover Publications, 2000) p. 298

7. K. Sabbagh, *The Riemann Hypothesis: The Greatest Unsolved Problem in Mathematics* (Farrar, Straus and Giroux, 2003) p. 188

8. A. Juhl, *Cohomological Theory of Dynamical Zeta Functions* (Birkhäuser, 2001)

9. K. Sabbagh, "Beautiful Mathematics", *Prospect* (January 2002) p. 43

10. M. Berry and J. Keating, "The Riemann Zeros and Eigenvalue Asymptotics", *SIAM Review* 41 (1999) p. 238

11. E. Klarreich, "Prime Time", *New Scientist* (11 November 2000) p. 33

12. M. du Sautoy, "The Music of the Primes", *Science Spectra* 11 (1998)

13. M. du Sautoy, *The Music of the Primes: Why an Unsolved Problem in Mathematics Matters* (HarperCollins, 2003) p. 58

14. *ibid*, p. 93

15. *ibid*, pp. 313–314

16. E. Klarreich, "Prime Time", *New Scientist* (11 November 2000) p. 36.

17. E. Landau, *Handbuch der Lehre von der Verteilung der Primzahlen* (Teubner, 1909)

18. I.J. Good, A.J. Mayne and J. Maynard Smith (eds.), *The Scientist Speculates* (Heinemann, 1962)

19. E.T. Bell, *Men of Mathematics* (Simon & Schuster, 1937) p. 538

20. K. Sabbagh, *The Riemann Hypothesis: The Greatest Unsolved Problem in Mathematics* (Farrar, Straus and Giroux, 2003) p. 36

21. J. Derbyshire, *Prime Obsession* (Joseph Henry Press, 2003)

22. D. Rockmore, *Stalking the Riemann Hypothesis: The Quest to Find the Hidden Law of Prime Numbers* (Pantheon Books, 2005)

23. R. Bellman, *A Brief Introduction to Theta Functions* (Holt, Rinehart & Winston, 1961) p. 33

24. M. du Sautoy, *The Music of the Primes: Why an Unsolved Problem in Mathematics Matters* (HarperCollins, 2003) p. 314

25. E. Klarreich, "Prime Time", *New Scientist* (11 November 2000) p. 35

26. H.M. Edwards, *Riemann's Zeta Function* (Dover Publications, 2000) p. 6

27. M. du Sautoy, "The Music of the Primes", *Science Spectra* 11 (1998)

28. K. Sabbagh, *The Riemann Hypothesis: The Greatest Unsolved Problem in Mathematics* (Farrar, Straus and Giroux, 2003) p. 36

29. M. Lapidus, *In Search of The Riemann Zeros: Strings, Fractal Membranes and Noncommutative Spacetimes* (AMS, 2008)

30. More accurately, what happens is that the prime counting function (familiar to us via its staircase graph) *evolves*, "morphing" into more general functions which generally won't be "step functions". As a result, rather than prime numbers maintaining their distinct identities but changing positions, we'd see them being "smeared" across the number line, in very particular ways.

31. B. Cipra, "A Prime Case of Chaos", *What's Happening in the Mathematical Sciences, Volume 4* (AMS, 1999) p. 17

32. K. Sabbagh, *The Riemann Hypothesis: The Greatest Unsolved Problem in Mathematics* (Farrar, Straus and Giroux, 2003) p. 268

33. B. Cipra, "A Prime Case of Chaos", *What's Happening in the Mathematical Sciences, Volume 4* (AMS, 1999) p. 5

34. E. Klarreich, "Prime Time", *New Scientist* (11 November 2000) p. 36

CHAPTER 26 [pages 189–196]

1. E. Klarreich, "Prime Time", *New Scientist* (11 November 2000) p. 36

2. K. Sabbagh, *The Riemann Hypothesis: The Greatest Unsolved Problem in Mathematics* (Farrar, Straus and Giroux, 2003), p. 268

3. M. du Sautoy, *The Music of the Primes: Why an Unsolved Problem in Mathematics Matters* (HarperCollins, 2003), p. 55

4. *ibid*, p. 18

5. H. Weyl, *Philosophy of Mathematics and Natural Science* (Princeton University Press, 1949) p. 7

6. K. Sabbagh, *The Riemann Hypothesis: The Greatest Unsolved Problem in Mathematics* (Farrar, Straus and Giroux, 2003) p. 22

7. M.C. Gutzwiller, *Chaos in Classical and Quantum Mechanics* (Springer, 1990) p. 308

8. K. Sabbagh, "Beautiful Mathematics", *Prospect* (January 2002) p. 40

9. M. du Sautoy, "The Music of the Primes", *Science Spectra* 11 (1998)

10. M. du Sautoy, *The Music of the Primes: Why an Unsolved Problem in Mathematics Matters* (HarperCollins, 2003) p. 5

11. *ibid*, p. 13

12. *ibid*, p. 18

13. E. Klarreich, "Prime Time", *New Scientist* (11 November 2000) p. 36

14. From publisher's description of J. Derbyshire, *Prime Obsession* (Joseph Henry Press, 2003).

15. M. du Sautoy, *The Music of the Primes: Why an Unsolved Problem in Mathematics Matters* (HarperCollins, 2003) p. 130

16. G. Tenenbaum, *Introduction to Analytic and Probabilistic Number Theory* (Cambridge University Press, 1995) p. 9

17. I. Stewart, *The Problems of Mathematics* (Oxford University Press, 1992) p. 164

CHAPTER 27 [pages 197–210]

1. A *Klein bottle* is a closed two-dimensional surface which is impossible to manifest in three-dimensional space, so you'll never actually see one, but it *is* possible to mathematically describe one. Imagine you have a square of very stretchy flat stuff. Imagine gluing two opposite edges together so that you now have a stretchy tube. Now look at each circular end of the tube. There are two ways you could travel around it, clockwise or anticlockwise. If we were to stretch the tube and then join up the ends to make a torus (the mathematical name for a "doughnut shape"), then we will have "joined clockwise to anticlockwise". Is there any other way we can join the ends up, so the join is "clockwise to clockwise"? In three dimensions there isn't, not without the tube actually piercing itself, and that's not allowed (the result wouldn't be a "closed two-dimensional surface"). But in higher dimensions it *is* possible to make this join, and the resulting closed (that is, "edge-free") two-dimensional surface is called a Klein bottle.

Similarly to the better known *Möbius strip*, which only has one side, the Klein bottle doesn't have an inside and an outside (like a spherical bubble does). If you imagine running around on the surface of a giant Klein bottle, you could find a path from any point back to that point, but so that your feet are were "underneath" where they started and your head was pointing away from the surface in the opposite direction. It's would be as if you'd gone from the outside to the inside, except that this distinction is meaningless because *there's only one side.*

2. A circle is considered to be a one-dimensional object (small pieces of it are similar to small pieces of line) which inhabits two-dimensional space (you draw circles on pieces of paper). A circle is defined as the set of points in a plane which are some fixed distance from the centre. If we take the same definition but replace "a plane" with "three-dimensional space", we get a set of points which make up a sphere. A sphere is considered to be a two-dimensional object (small pieces of a sphere are similar to small pieces of a plane) living in three-dimensional space.

Now suppose we continue with this analogy. We want to find a comparable three-dimensional object living in four-dimensional space. Without having to get confused about Einstein, space-time continuums, *etc.*, we can introduce a fourth dimension to our geometry mathematically. A "hypersphere" in four dimensions is the set of points some fixed distance from a fixed point. The concept of "distance" generalises easily to four (or higher) dimensions, so we can keep doing this all the way up.

In terms of equations, it looks like this (we're taking the radius to be 1 here and the centre to be the "origin" of the coordinate system – where all axes cross and all coordinates are zero):

Circle in plane [(x,y) coordinate space]:

$$x^2 + y^2 = 1$$

Sphere in three-dimensional space [(x,y,z) coordinate space]:

$$x^2 + y^2 + z^2 = 1$$

Hypersphere in four-dimensional space [(w,x,y,z) coordinate space]:

$$w^2 + x^2 + y^2 + z^2 = 1$$

Hypersphere in five-dimensional space [(v,w,x,y,z) coordinate space]:

$$v^2 + w^2 + x^2 + y^2 + z^2 = 1$$

...and so on.

3. You may have heard of the "wave-particle duality" in quantum physics. The idea is that light (or any other form of electromagnetic energy) can be thought of either as a stream of particles or as a wave, depending on how you choose to observe it. Philosophically speaking, there's still not an entirely satisfactory explanation for this, but physicists have learned to work with it over the decades. Around 1900, Max Planck related the energy of a particle to the frequency of the associated wave as follows:

$$h = E\nu,$$

where *h* is the *Planck constant*, *E* is energy and ν is frequency. Wavelength λ and frequency ν are related as follows, where *c* is the speed of light:

$$\lambda = c/\nu.$$

4. A 5 × 5 matrix won't always have 5 *distinct* eigenvalues, as sometimes values can be repeated.

5. B. Cipra, "A Prime Case of Chaos", *What's Happening in the Mathematical Sciences*, Volume 4 (AMS, 1999) p. 16

6. M. du Sautoy, *The Music of the Primes: Why an Unsolved Problem in Mathematics Matters* (HarperCollins, 2003) p. 280

Appendix 10

Euler's product formula

Let $s = x + iy$, so $\mathrm{Re}(s) > 1$ means that $x > 1$, which means that for any prime p, $p^x > 1$. Using the notation $|z|$ for the modulus of a complex number z, it follows that

$$|1/p^s| = |p^{-s}| = |p^{-x-iy}| = |p^{-x}p^{-iy}| = |p^{-x}||e^{-iy\log p}| = |p^{-x}| \cdot 1 = 1/|p^x| < 1$$

A well-known result on convergent series tells us that for $|z| < 1$,

$$\frac{1}{1-z} = 1 + z + z^2 + z^3 + z^4 + \cdots$$

(multiply both sides by $1-z$ to see this). So, in the half-plane to the right of the critical strip, where $\mathrm{Re}(s) > 1$, it follows that

$$\frac{1}{1-p^{-s}} = 1 + \frac{1}{p^s} + \left(\frac{1}{p^s}\right)^2 + \left(\frac{1}{p^s}\right)^3 + \left(\frac{1}{p^s}\right)^4 + \cdots$$
$$= 1 + \frac{1}{p^s} + \frac{1}{p^{2s}} + \frac{1}{p^{3s}} + \frac{1}{p^{4s}} + \cdots$$

We can then write out the infinite product from one side of Euler's product formula,

$$\prod_p \frac{1}{1-p^{-s}} = \left(\frac{1}{1-2^{-s}}\right)\left(\frac{1}{1-3^{-s}}\right)\left(\frac{1}{1-5^{-s}}\right)\cdots$$

as

$$\left(1 + \frac{1}{2^s} + \frac{1}{2^{2s}} + \frac{1}{2^{3s}} + \cdots\right)\left(1 + \frac{1}{3^s} + \frac{1}{3^{2s}} + \frac{1}{3^{3s}} + \cdots\right)\left(1 + \frac{1}{5^s} + \frac{1}{5^{2s}} + \frac{1}{5^{3s}} + \cdots\right)\cdots$$

Because this product is known to converge, each term in the resulting sum can be seen to be built from a finite number of factors drawn from the bracketed contents associated with each prime:

$$1 + \frac{1}{2^s} + \frac{1}{3^s} + \frac{1}{2^{2s}} + \frac{1}{5^s} + \frac{1}{2^s}\cdot\frac{1}{3^s} +$$
$$\frac{1}{7^s} + \frac{1}{2^{3s}} + \frac{1}{3^{2s}} + \frac{1}{2^s}\cdot\frac{1}{5^s} + \frac{1}{11^s} + \cdots$$

We see that every finite combination of primes (to the power $-s$) appears exactly once in this expansion. By the Fundamental Theorem of Arithmetic, this means that each counting number (to the power $-s$) will appear exactly once (as each counting number is guaranteed to have a unique prime factorisation). This means that our expansion reduces to

$$1 + \frac{1}{2^s} + \frac{1}{3^s} + \frac{1}{4^s} + \frac{1}{5^s} + \cdots + \frac{1}{n^s} + \cdots = \sum_{n=1}^{\infty} \frac{1}{n^s}$$

as required.

appendix 11

why there are no nontrivial zeta zeros outside the critical strip

In his original paper, Riemann observed that $\zeta(s)$ has no zeros in $\text{Re}(s) > 1$ because in that half-plane it can be represented as

$$\prod_p \frac{1}{1-p^{-s}} = \left(\frac{1}{1-2^{-s}}\right)\left(\frac{1}{1-3^{-s}}\right)\left(\frac{1}{1-5^{-s}}\right)\cdots$$

and a convergent infinite product (which this is) can only equal zero if one of its factors is zero – and none of these factors can be zero, clearly.

To the *left* of the critical strip, that is, when $\text{Re}(s) < 0$, we can use the functional equation (see Appendix 12) to relate the behaviour of $\zeta(s)$ to its behaviour on the *right* of the critical strip via the relation $\Lambda(1-s) = \Lambda(s)$, where $\Lambda(s) = \Gamma(s/2)\zeta(s)\pi^{-s/2}$. We know that $\zeta(s)$ is nonzero to the right of the strip. Also, all powers of π (real or complex) must be nonzero and the function Γ has no zeros anywhere. As $\Lambda(s)$ is nonzero for $\text{Re}(s) > 1$, we see that $\Lambda(1-s)$ cannot be zero for $\text{Re}(1-s) < 0$.

This means that we can be sure $\zeta(s)$ is nonzero to the left of the critical strip *as long as* $\Gamma(s/2)$ *is defined*. It turns out that $\Gamma(s)$ has poles at $-1, -2, -3, -4$, etc., so the points we have to watch out for are $-2, -4, -6, -8...$, which we already know to be the trivial zeros of $\zeta(s)$. Notice that if it weren't for the poles of the gamma function (something like infinite values which it takes at these isolated points), the functional equation would tell us that the zeta function should have zeros at $1-(-2)=3$, $1-(-4)=5$, $1-(-6)=7$, *etc.*, which we know it doesn't.

We can thus conclude that there are no *nontrivial* zeros of the Riemann zeta function outside the critical strip.

appendix 12

the functional equation of the Riemann zeta function

Here's the equation again:

$$\frac{\Gamma(\frac{s}{2})\zeta(s)}{\pi^{s/2}} = \frac{\Gamma(\frac{1-s}{2})\zeta(1-s)}{\pi^{(1-s)/2}}$$

s represents a typical complex number (most easily thought of as a point in the complex plane), and $1-s$ is the complex number which, graphically, can be thought of as the point s reflected across the point $1/2$ (see page 73).

$\zeta(s)$ represents the usual Riemann zeta function applied to the complex number s. π is just the familiar 3.14159..., although you'll notice that here it's being raised to a complex number power on both sides of the equation. As we've seen, such a power can be understood in terms of spirals, although there are other approaches to understanding $\pi^{s/2}$. If we rewrite π as $e^{\log \pi}$, this becomes $e^{(s/2)\log \pi}$. Since $s/2$ can be split into $x/2 + iy/2$, where x and y are the respective real and imaginary parts of s, this becomes $e^{(x/2)\log \pi} e^{i(y/2)\log \pi}$. The first factor is just e raised to a real power, whereas the second can be further expanded using the trigonometric identity $e^{i\theta} = \cos\theta + i\sin\theta$.

Γ here represents the *gamma function*. This is an extension of the more familiar *factorial function* $n! = n \cdot (n-1) \cdot (n-2) \cdot 2 \cdot 1$ to complex numbers. For reasons of convention, $\Gamma(n) = (n-1)!$, rather than the more obvious $n!$ Initially, it's not at all clear how a function like the factorial can be extended to noninteger reals, let alone complex numbers. But if, for a complex number z such that $\text{Re}(z) > 0$, we define the gamma function in terms of the integral

$$\Gamma(z) = \int_0^\infty t^{z-1} e^{-t} dt$$

then it's not too difficult to show that when z equals an integer, $\Gamma(z) = (z-1)!$ This integral doesn't give finite values when z is taken in the left half-plane, but the function it defines on the right half-plane can be *analytically continued*[1] to all of $\mathbb{C}$ except the negative integers.

So the functional equation of the Riemann zeta function tells us that if we define the function $\Lambda(s) = \Gamma(s)\zeta(s)/\pi^{s/2}$, then it satisfies the relation $\Lambda(s) = \Lambda(1-s)$. Equivalently, this function is symmetric with respect to reflection across the point $s = 1/2$.

The discovery of the functional equation is usually attributed to Riemann, although in his comprehensive book on the zeta function, Harold Edwards mentions that Euler gave the equation in a slightly different form in a 1761 paper[2], and that "it is entirely possible that Riemann found it there"[3]. In any case, Riemann was certainly the first to prove it. His 1859 paper actually includes *two* proofs of the functional equation, both of which are explained in great detail in Edwards' book.

notes

1. Back in Chapter 18, I described how Euler's zeta function, defined on part of the real number line, could be extended to the complex plane in a "uniquely natural" way, giving rise to a complex function with the property of being "analytic". This is analytic continuation, but to explain it properly would require another lengthy appendix.

2. L. Euler, "Remarques sur un beau rapport entre les séries des puissances tant directes que réciproques", *Mem. Acad. Sci. Berlin* 17 (1761) pp. 83-106. Edwards mentions this in a footnote on his page 12.

3. H.M. Edwards, *Riemann's Zeta Function* (Dover Publications, 2000)

appendix 13

the behaviour of the Riemann zeta function in the "critical strip"

As we've seen, we can calculate what the Riemann zeta function does to the right of the critical strip $0 < \text{Re}(s) < 1$ using either an infinite sum or an infinite product. The functional equation (see Appendix 12) then allows us to deduce what it does to the left of the strip. We know that there's an extension to the whole of the complex plane (apart from the pole $s = 1$), but how can we determine the behaviour of zeta *within* the critical strip?

This requires a trick involving the *Dirichlet eta function*:

$$\eta(s) = \frac{1}{1^s} - \frac{1}{2^s} + \frac{1}{3^s} - \frac{1}{4^s} + \cdots$$

The eta function (named after another Greek letter) is sometimes called the "alternating zeta function", for reasons which should be obvious. Unlike the infinite sum associated with the zeta function itself, which only converges for $\text{Re}(s) > 1$, the above infinite sum helpfully converges for the slightly larger region $\text{Re}(s) > 0$ (note that it involves terms being both added *and* subtracted, which might give you some feeling as to why this is).

This means that

$$\zeta(s) - \eta(s) = \left(\frac{1}{1^s} + \frac{1}{2^s} + \frac{1}{3^s} + \frac{1}{4^s} + \cdots\right) - \left(\frac{1}{1^s} - \frac{1}{2^s} + \frac{1}{3^s} - \frac{1}{4^s} + \cdots\right)$$

$$= 2\left(\frac{1}{2^s} + \frac{1}{4^s} + \frac{1}{6^s} + \frac{1}{8^s} + \cdots\right) = \frac{2}{2^s}\left(1 + \frac{1}{2^s} + \frac{1}{3^s} + \frac{1}{4^s} + \cdots\right)$$

so we have

$$\eta(s) = \zeta(s) - 2^{1-s}\zeta(s) = (1 - 2^{1-s})\zeta(s)$$

and so

$$\zeta(s) = \left(\frac{1}{1-2^{1-s}}\right)\eta(s) = \left(\frac{1}{1-2^{1-s}}\right)\left(\frac{1}{1^s} - \frac{1}{2^s} + \frac{1}{3^s} - \frac{1}{4^s} + \cdots\right)$$

This allows values of the Riemann zeta function to be calculated within the critical strip. With this and the functional equation at our disposal, we now have the means to compute the value of zeta anywhere on the complex plane (except the pole at 1, of course).

Although mathematically satisfying and entirely valid, these expressions for the zeta function involving infinite sums are usually not the most efficient means to calculate its values. More sophisticated algorithms for such calculations are being developed all the time, particularly in regard to locating new zeros of the function.

appendix 14

precise formulation of the Prime Number Theorem

In the usual mathematical notation, the PNT tells us that both

$$\pi(x) \sim x/\log x \quad \text{and} \quad \psi(x) \sim x$$

where "$\pi(x)$" is the number of primes less than or equal to x, "$\psi(x)$" is the amount of "primeness"[1] less than x, and "$x/\log x$" is the number x divided by its own natural logarithm. The ~ symbol ("tilde" or "twiddle") here indicates a particular form of approximate equality, in this case telling us that the ratios

$$\frac{\pi(x)}{x/\log x} \quad \text{and} \quad \frac{\psi(x)}{x}$$

can be brought as close to 1 as you require (and will stay at least this close) by taking x to be sufficiently large. More precisely, it means that if you are given *any* (usually small) positive real number ε, then there will always be some (usually large) positive number k such that, for all x greater than k,

$$\left|\frac{\pi(x)}{x/\log x} - 1\right| < \epsilon \quad \text{and} \quad \left|\frac{\psi(x)}{x} - 1\right| < \epsilon$$

notes

1. Remember, this is a sum of logarithms of primes (counting both primes and their powers). See Chapter 12 for a reminder of the exact definition.

Index

2001: A Space Odyssey 121

A Beautiful Mind 180
addition explained graphically in $\mathbb{C}$ 18
addition explained graphically in $\mathbb{R}$ 12
aerodynamics 26
American Institute of Mathematics 179
analytic continuation 230
analytic number theory 46, 116, 125, 126, 179
Archbishop of Canterbury 124
Archimedes 124

Backlund, Ralf J. 107, 109
Barnes, Ernest 179
bell curve 126–30, 133–5, 137
Bell, Eric Temple 141, 180
Bellman, Richard 119, 140, 181
Berry, Michael 174–5, 182
Big Bang 124
Bombieri, Enrico iv, 99, 112, 113, 141, 142–3, 174, 186
Brent, Richard P. 109
Buddhism 135–7

$\mathbb{C}$ (system of complex numbers) 15–28
   functions on 30–6
Cambridge University 108, 179
categories (of named things) 135–7
Central Limit Theorem 134
central limit theorems 134–5
Chladni, Ernst, 201
Chladni figures 201–2
Cipra, Barry 186
Clay Mathematics Institute 180
commutative operations 13
completion of a number system 11, 212–13
complex analysis 25–6
complex plane 15–23, 25, 27–8
   functions on 30–6
Connes, Alain 146
Conrey, J. Brian 106, 144, 178
   "40%" result on nontrivial zeta zeros 146, 171
constructivism, mathematical 213

Crick, Francis 124
cryptography 189

Darwin, Charles 124
de la Vallée Poussin, Charles-Jean 1, 78, 116, 178
Deitmar, Anton 171
Deligne, Pierre, 220
Derbyshire, John 180, 181, 183, 194
Descartes, René 124
deviation in "primeness" count 1, 2
   as "error" in PNT 114–6, 142
   decomposition into spiral waves 81, 84, 112–14, 174
   rate of growth 151, 156
Devlin, Keith, 141
Dirichlet eta function 231
division, explained graphically in $\mathbb{R}$ 14, 44–5
du Sautoy, Marcus 141, 167, 175, 181, 182, 191, 193, 194, 210
   The Music of the Primes 173, 180, 183
dynamical systems 186, 204–5, 208
   (spectra of) energy levels 205

e 11, 229
   base-e spiral 57, 214
   irrationality, Fourier's proof of 212
Edwards, Harold M. 172, 182, 230
eigenvalues 203, 207–10
Einstein, Albert 124
electricity 26
electromagnetic spectrum 204
Elon, Amos 120
enculturation 136
Enriques surface 204
Erdős, Paul 127
Erdős–Kac Theorem 127, 130–5
Euclid 116–7, 178
Euler, Leonhard 36–7, 42, 43, 45
   calculating values of his zeta function 107
   extending FTA to "analytic" context 117
   giving version of functional equation 230
   proving product formula 46
Euler–Maclaurin summation formula 107

Euler product formula 46–7
  and absence of zeros to right of critical strip 215, 228
  as "extension" of FTA 116–17, 149
  description of proof 226–7
Euler zeta function 36–50, 53–6, 61, 100, 214
  extension to $\mathbb{C}$ (minus one point) 49–50, 53–5, 65, 79, 100
  graph of 41
  lack of zeros 47, 54, 65
explicit formula, Riemann's 82, 84–5, 97, 99, 113
  and PNT 116
  role of zeros in 99
  significance of 104
  "turned inside out" 101–2

factorial function 215, 229
factorisation, unique 46, 149, 227
Farey, John, Sr., 220
Farey series 162–6
Fermat's Last Theorem 182, 185
fixed points (of a function) 30, 34
flow 185–6
Fourier analysis 102, 216
Fourier duality 102, 216
fractals 27–8
fractions 8–9
functions 3, 29–30
  bounding 151, 154
  complex 29–36
    analytic (or holomorphic) 50
  continuous 34, 52, 61, 66, 68
  Euler zeta: see *Euler zeta function*
  families of 34, 50, 67, 170–1
  fixed points of: see *fixed points*
  graph of: see *graph*
  L-: see *L-functions*
  Riemann zeta: see *Riemann zeta function*
  zeros of 34–6, 54, 67
Fundamental Theorem of Arithmetic 149
  and Euler's product formula 46–7, 116–17, 149, 227
  Euclid's discovery of 117

Galton, Francis 125, 127, 217
gamma function 75, 215, 228, 229
Gauss, Carl, 175, 178, 218
Gaussian distribution 126–30, 134–5, 137, 218
GIMPS (Great Internet Mersenne Prime Search) 110
glamour 120
Gowers, Timothy 130
Gram, Jørgen P. 107, 109

graph (of a function) 29
  illustrating zeros of a function 35
  impossibility in context of ζ 33, 51
Gutzwiller, Martin C. 119, 193

Hadamard, Jacques 1, 78, 116, 178
Hardy, Godfrey
  remark on probability 134
  work on Riemann zeta function 106, 144, 179
  work with Ramanujan 127, 131
Hardy–Ramanujan Theorem 131
Hasse, Helmut 220
Hawking, Stephen 124
Hilbert, David 140, 181, 182, 191
Hinduism 125, 218
histogram 128–30, 133, 134
Holmes, Sherlock 122
Hutchinson, John I. 107, 109
Huxley, Martin 141
hypersphere 204, 223–4

i (the imaginary unit) 16–17
  multiplied by itself 22
imaginary numbers 17
imaginary part (of a complex number) 18
infinite products
  of complex numbers 64, 226
  of real numbers 44
infinite sums
  of complex numbers 59, 226–7, 231–2
  of real numbers 36–8, 40
  of spiral waves 84
integers 6–7
  negative 6
inverse spectral problems 203
irrational numbers 10–11
Ivić, Aleksandar 143
Iwaniec, Henryk 143, 183, 190

Jerusalem 120
jñeyāvarana 136–7
Juhl, Andreas 172
Jung, Carl 120
Jutila, Matti 173

Kac, Mark 127
Keating, Jon, 174
Klarreich, Erica 174, 178, 193–4
Klein bottle 204, 223

L-functions 34, 170–2, 220
Lakshmi Namagiri 125, 127
Landau, Edmund 179
Lapidus, Michel 185–6
Law of Large Numbers 134
limit 46, 133–4, 213, 214
Littlewood, John 179, 194
  opinion that RH is probably not true 179
  remark on probability 134
  work on Riemann zeta function 106, 144, 179

Maslanka, Krzysztof 119
matrices 206–10
  algebra of 206
  entries of 206
  (spectra of) eigenvalues of 207
Mendès France, Michel 103
Mertens function 157–61
Möbius strip 223
Montgomery, Hugh, 141, 180–1
Motohashi, Yoichi 186, 193
multiplication explained graphically in $\mathbb{C}$ 19–23
multiplication explained graphically in $\mathbb{R}$ 13
musical analogies 97, 99, 104, 124, 172–7, 181
mystery, etymology of word 122

$\mathbb{N}$ (natural numbers or counting numbers) 6–7
Namagiri 125, 127
Nash, John 180
negative integers 6
negative numbers 6–7
  multiplying 23–5
"neo-Pythagoreanism" 195
Newton, Isaac 124
number systems 5–28

Odlyzko, Andrew 110, 144, 210
operators 209
  (spectra of) eigenvalues of 209–10

p-adic number systems 11, 213
Peano axioms 147
Penrose, Roger 213
π (pi) 10, 48, 66, 94, 228, 229
Planck constant 224
Platonism 130, 213
pole of Riemann zeta function 52, 72, 85–6
poles of general complex functions 66
"power curves" 153–6, 161, 166

powers 152
prapañca 136
prime factorisation 157–61, 225
  and probability 48, 130–33, 134
prime numbers 1–3, 11, 117
  and Euler zeta function 43, 47, 100
  and explicit formula 84
  and Riemann zeta function 62–4
  distribution of 1–2, 79, 81
    and Riemann Hypothesis 142–4, 147–9
    as built from spiral waves 97, 112–13, 142–3
    harmonic decomposition 99, 174
  relationship with Riemann zeros 99–104
Prime Number Theorem
  and location of zeta zeros 78–9, 116, 139
  "error" in 114–15, 142
  precise formulation 233
  proof of 1, 78, 116
primeness counting function 82–5, 97
probabilistic number theory 127
probability 134–5
  of finding relatively prime integers 48, 66
Pythagoras 10
Pythagorean theorem 10
Pythagoreanism, 195

$\mathbb{Q}$ (system of rational numbers) 9–11, 162–5, 213
  "holes" in 9–11
quadrivium 176
qualitative, quantitative approaches to number 119, 126
quantum (subatomic) physics 26, 124, 205, 207, 208, 224
quest 181

$\mathbb{R}$ (system of real numbers) 11–12, 15
Ramanujan, Srinivasa 125, 127, 131
randomness 127
  associated with number system 130–31, 144, 167–8
real part (of a complex number) 18
relatively prime numbers 48, 66
Riemann, Bernhard 178
  explicit formula: see *explicit formula, Riemann's*
  introduction of zeta function 1–2, 79, 117
    harmonic decomposition 104, 156, 175
    work on zeros 77, 78, 107
    speculation (RH) 77, 139, 169, 179
Riemann Hypothesis 139–49
  and relationship between addition and multiplication 146–9
  claimed proofs and disproofs 183–4
  classical 170

Riemann Hypothesis (*continued*)
  generalised 170
  Grand 170
  reformulations of 146, 151–68
  significance of 140–2, 169–86, 189–96
Riemann zeta function 2–3, 29, 119–20, 170
  continuity of 52, 66
  critical line 71, 76–7, 105–6
  critical strip 69–77, 104, 139
    zero-free region 114–15
  functional equation of 72–5, 228, 229
  images generated from 51
  pole of 52, 72, 85–6
  possible "religious" significance 121–5
  significance of point ½ 72, 78–9
  zeros of 54–5, 65, 68, 100
    nontrivial 69–71, 85–7, 103–106
      "99% close to critical line" result 145–6
      calculations of 107–10
      counting function 101–2
      spectral interpretation of 197, 210
      why all in critical strip 228
    trivial 69, 75, 85–6, 104
Rockmore, Dan 181, 183

Sabbagh, Karl 180, 183, 193
Sarnak, Peter 72, 112, 121, 141, 186, 190, 210
Selberg, Atle, 172, 186
sine waves
  compared to spiral waves 82, 87–9, 91–2
  decomposition into 102, 174, 198–9
Smith, Warren D. 120
spectra 197–210
spectral analysis 199, 203
spectral geometry 201, 203
spectroscopes 198
spirals, logarithmic (or equiangular)
  base of 38
  constructing spiral waves via 93–6
  defining Euler zeta function via 38–40, 43–44
  defining Riemann zeta function via 56–62
  defining square root via 132
spiral waves 1, 3
  relationship to prime distribution 112–14, 143–4, 175
  relationship to zeta zeros 81–98
  "specifications" 82, 87–98
    frequency 88, 91
    phase point 92–5, 98
    rate of amplitude growth 89–97, 110–12, 142–3

square roots 10, 132
standing waves 200–2
starlight 198, 208
statistical independence 135
Stewart, Ian 194
subtraction explained graphically in $\mathbb{C}$ 19
subtraction explained graphically in $\mathbb{R}$ 12

Tenenbaum, Gérald 103, 194
Tibetan prayer bowls 199
transfinite numbers 212
Trinity College, Cambridge 108, 179
Turing, Alan 108–9, 217
Turing machines 217

unit 6

von Lindemann, Ferdinand 212

Watson, James D. 124
wave–particle duality 224
Wedeniwski, Sebastian 109, 110
Weil, André 220
Weyl, Hermann 193
white noise 199
Wiles, Andrew 182, 184
Winnie-the-Pooh 52, 214

$\mathbb{Z}$ (system of integers) 7
Zagier, Don 193
zeros (of a function) 34–6, 54, 67
zeta functions, other 170–2
ZetaGrid 110, 144